W9-BTO-277

HOMEMADE
LIVING

Keeping Bees

with Ashley English

Keeping Bees
with Ashley English

All You Need to Know to Tend Hives, Harvest Honey & More

LARK
CRAFTS
An Imprint of Sterling Publishing Co., Inc.
New York

WWW.LARKCRAFTS.COM

Senior Editor: Nicole McConville

Editorial Assistant: Beth Sweet

Creative Director: Chris Bryant

Layout & Design: Eric Stevens

Illustrations: Eric Stevens and
Melanie Powell

Photographer: Lynne Harty

Additional Photography: Steve Mann

Cover Designer: Eric Stevens

Library of Congress Cataloging-in-Publication Data

English, Ashley, 1976-
  Homemade living : keeping bees with Ashley English : all you need to know to tend hives, harvest honey & more / Ashley English.
    p. cm.
  Includes index.
  ISBN 978-1-60059-626-1 (hc-plc : alk. paper)
  1. Bee culture. I. Title.
  SF523.E54 2011
  638'.1--dc22

                2010020674

10 9 8 7 6 5 4 3 2 1

First Edition

Published by Lark Crafts
An Imprint of Sterling Publishing Co., Inc.
387 Park Avenue South, New York, NY 10016

Text © 2011, Ashley English
Photography © 2011, Lark Crafts, an Imprint of Sterling Publishing Co., Inc., unless otherwise specified
Illustrations © 2011, Lark Crafts, an Imprint of Sterling Publishing Co., Inc., unless otherwise specified

Distributed in Canada by Sterling Publishing,
c/o Canadian Manda Group, 165 Dufferin Street
Toronto, Ontario, Canada M6K 3H6

Distributed in the United Kingdom by GMC Distribution Services,
Castle Place, 166 High Street, Lewes, East Sussex, England BN7 1XU

Distributed in Australia by Capricorn Link (Australia) Pty Ltd.,
P.O. Box 704, Windsor, NSW 2756 Australia

If you have questions or comments about this book, please contact:
Lark Crafts
67 Broadway
Asheville, NC 28801
828-253-0467

Manufactured in Canada

ISBN 13: 978-1-60059-626-1

For information about custom editions, special sales, premium and corporate purchases, please contact Sterling Special Sales Department at 800-805-5489 or specialsales@sterlingpub.com.

For information about desk and examination copies available to college and university professors, requests must be submitted to academic@larkbooks.com.
Our complete policy can be found at www.larkcrafts.com.

 This book was printed on recycled paper with agri-based inks.

Table of Contents

INTRODUCTION 8

Chapter 1: UNDERSTANDING BEES 12

Humans & Honeybees 13

Honeybee Anatomy 101 14

Hive Hierarchy 18

Becoming Adults 21

Hive Talkin' 22

Chapter 2: WHAT TO CONSIDER 24

On Location 25

Money Matters 25

Keeping Time 26

Law-Abiding Citizen 27

Being Neighborly 27

Pets & Children 28

Allergies 29

Species Variations 30

Chapter 3: HOUSING 32

Anatomy of a Hive 33

Hive Components 33

Where to Place Your Hive 40

Setting Up House 41

Chapter 4: FEEDING BEES 42

The Birds & the Bees 43

A Need for Feed 43

Feed Options 44

Feeders 49

Chapter 5: ESSENTIAL EQUIPMENT 50

Smoker & Fuel 51

Suits & Veils 52

Gloves 53

Hive Tool 53

Helpful Extras 54

Chapter 6: OBTAINING BEES 56

A Time & a Place 57

A Package Deal (Package Bees) 57

Installing a Package 59

(Nuc)lear Energy 62

Swarming into Action 62

The Established Order 64

Supply & Demand 65

Chapter 7: A LOOK INSIDE 66

Being a Good Houseguest 67

Inspector General 67

Opening the Hive 68

Removing Frames 69

What to Look For 71

Replacing Frames 72

Closing the Hive 73

Chapter 8: A YEAR OF BEES 74

Spring 75

Summer 80

Autumn 81

Winter 82

Chapter 9: HEALTH & WELLNESS 84

Diseases 85

Parasites 87

Pests 90

Colony Collapse Disorder 94

Chapter 10: HONEY 96

Making Magic 97

The Honey House 98

The Harvest 98

Extraction Action 106

The Honey Larder 109

Honey Money 110

What's in a Name 110

Sweet Rewards 111

Health from the Hive 112

Chapter 11: HONEY RECIPES 114

Infused Honey 115

Holiday Rounds 116

Grain Mustard Honey Vinaigrette 118

Roasted Root Vegetables with Honey & Herbs 119

Chestnut Soup with Honey 120

Honeyed Prawns & Polenta 122

Honey Ice Cream 124

Fruit, Nut & Honey Granola 125

Hot Cider & Honey Toddy 126

Honey & Ginger Cold-Fighting Tea 127

NOTES 128

RESOURCES 129

GLOSSARY 130

ACKNOWLEDGMENTS 133

INDEX 134

Introduction

From rolling fields to rooftops, a food renaissance is taking root. In community gardens, grade school playgrounds, and both country and urban kitchens, everyone from young children to sage grandmothers are finding pleasure embracing the DIY food movement, giddy with excitement over the thought of growing an abundance of backyard produce, putting up batches of fresh strawberry jam, or transforming milk into tender mozzarella.

Whether you're already onboard or merely ready to set sail, you're in good company, amongst kindred spirits. What brought you here matters less than where you'll end up. From folks wanting clean, quality foods to offer their families, to those interested in eliminating the culinary middleman (and saving money along the way), to those simply wanting to move a bit closer to their food's point of origin, the emerging food movement crosses all divides and covers all bases.

Keeping a backyard (or even rooftop!) beehive is a fantastic step toward championing homegrown foodways. When you consider that over one-third of the foods eaten by humans (including foods consumed by the animals that humans rely on) depend in some part on the fastidious efforts of honeybees, the importance of keeping a hive, or several, takes on deeper significance. Taking on the stewardship of honeybees benefits our food supply in countless ways. In fact, I know several beekeepers who keep hives but never even extract honey.

My interest in keeping bees started because I wanted to ramp up the production of my vegetable garden. I'd heard from a beekeeping friend that her garden's output had more than doubled since introducing honeybees to her yard. That sounded good to me. But I also should confess that ever since reading Sue Monk Kidd's *The Secret Life of Bees* several years back, I've been enchanted by the winged beauties, curious to learn their secrets and study their intrinsically egalitarian ways. In a hive, every bee's role is crucial; everyone matters. Honeybees live in absolute symbiosis, with each task, each role promoting the continuation of the hive and helping it to flourish. "I want to be privy to that world," I'd find myself thinking. Even then, I sensed that, while I might become a so-called "beekeeper," it was really I who would be kept, captivated and enthralled by the bees.

Once I decided that the beekeeping life was for me, I set about finding a community of fellow bee lovers. It was easy. I believe you'll find yourself pleasantly surprised at the number of beekeepers in your community. For me, the best part (even better than the bee club field days, informative monthly meetings, and vast pools of knowledge and expertise available from the beekeepers in my neck of the woods) is the opportunity to rub shoulders with people I might otherwise never encounter. While we might run in completely different social and cultural worlds, come beekeeping time, all differences are put aside. We're united in an overarching kinship of honeybee care and concern.

In this book, I share with you all the nitty-gritty details and learned-from-experience tips I've discovered in my own beekeeping journey. I examine the questions to ask and concerns to consider well in advance of purchasing your first package of bees; discuss hive hierarchy and exactly what "bee space" is; offer a thorough examination of hive housing, location, and feeding requirements; detail obtaining and installing honeybees; and cover how to extract honey. You'll also find absolutely mouth-watering honey-based recipes, concocted and perfected in my home kitchen; profiles of beekeepers in a variety of professions and trades; and essential information on how best to provide for the health and wellness of your hives.

My sincere hope is that this reference will serve as a useful and enjoyable companion to you along the way, giving you the guidance I really wish I'd had when I first got started. I took the plunge into this new way of living with my eyes wide open, and so can you. Precious little space is needed to get going; in fact, you don't even need a yard! For me, keeping bees has been more than just a hobby; it's been truly life changing. Witnessing the silent cooperation that goes on within a hive and the profoundly vital task of pollination has given me an even greater appreciation for and commitment to the miraculous cycle of life unfolding around us daily.

I invite you to come along for the ride. Together we'll examine all of the steps necessary to successfully take on stewardship of 50,000 creatures. I'll cheer quietly from the sidelines, helping you actualize your own apiary longings. You might get stung, you might get nervous, you might spill some honey. No bother. You'll also find the entire experience exhilarating beyond description.

Ashley English

ABOUT THE AUTHOR

Years ago, I was hopping into my car each morning, heading off to a job in a medical office. Things changed, though, when a whirlwind romance quickly resulted in marriage, a little homestead at the end of a dirt road, and just the encouragement and support I needed to make some serious life changes. Combining my long-standing interest and education in nutrition, sustainability, and local food, I made the bold decision to leave my stable office job and try my hand at homesteading. It was a huge leap of faith, but I truly believed there was opportunity waiting in a simpler, pared-down life. My goal was to find ways to nourish both body and soul through mindful food practices. And so I jumped in, rubber boots first, completely unaware of what lay ahead.

In my desire to chronicle both the triumphs and lessons of crafting a homemade life, I started up a blog, Small Measure (www.small-measure.blogspot.com). In it, I try to convey the same ideals I live every day: there are small, simple measures you can take to enhance your life while also caring for your family, community, and the larger world. It's been a trial-and-error experiment in living, full of a few pitfalls along with the joy. I've learned so much along the way, and I hope this book serves as continual encouragement for you. If I did it, you certainly can, too.

Chapter 1
Understanding Bees

Humans have had bees in their bonnets, as it were, since the first hunter-gatherer sat beside a buzzing log and got a whiff of the ambrosial fragrance emanating from within. Honeybees are one of the few types of insects purposely kept and managed by humans (observatory ant farms or flea circuses notwithstanding). Long admired for their social organization, architectural feats, and, of course, their products, honeybees are fascinating creatures in their own right. While most beekeepers are in it for the honey, maintaining a thriving hive requires a thorough understanding of bee behaviors, preferences, quirks, and basic make-up. Fledgling beekeepers will benefit enormously from learning a bit of honeybee anatomy, social organization, communication, and life stages before lighting up the smoker or donning a bee veil.

HUMANS & HONEYBEES

Prehistoric cave paintings in Africa and Western Europe, dating as far back as 15,000 B.C., depict scenes of harvesting honey and beeswax from wild bee colonies. Before the invention of specialized units for housing bees, gatherers would climb trees and fill wild beehives with smoke in order to harvest honey. Our intrepid ancestors undoubtedly subjected themselves to countless bee stings along the way, as protective bee clothing is a modern phenomenon. Until sugarcane made its way out of the tropics and around the world, honey was the primary sweetener available, so coveted that it was worth all the suffering required to obtain it.

Ancient Egyptian beekeepers were hive pioneers, housing honeybees in stacked cylinders made of clay, dung, or woven grasses. Ancient Rome and Greece engaged in beekeeping, as well, housing bees in a variety of structures, including hollowed-out logs, pottery, and straw baskets called skeps, the form that still comes to mind to many when they hear the word "beehive." Early Northern Europeans first harvested honey from wild hives, using systems of ropes, slings, pick axes, and even spiked footwear for scaling trees. They eventually moved on to use of structures similar to those used by ancient Greeks and Romans.

Unfortunately, the design flaw of these early housing units was that in order to harvest honey or beeswax—or even to take a peek inside—the housing unit had to be destroyed, along with the entire hive. With few exceptions, this remained the case until the invention of the moveable hive by Lorenzo Langstroth in 1852. A pastor from Pennsylvania, Langstroth created a boxlike hive structure composed of interchangeable, easily removable frames. Langstroth's keen understanding of *bee space*, or how much room honeybees (workers, drones, and queen) require to move about easily within a hive, was essential to the success of the design, which has survived, virtually unaltered, for more than 150 years. No small wonder then that he is often considered the father of American beekeeping.

A collection of antique bee skeps

HONEYBEE ANATOMY 101

The body parts of bees enable these versatile creatures to execute a large and varied set of tasks. Gaining a basic understanding of bee anatomy will prove beneficial not just for knowing what's going on and who's doing what, but for learning how to tell if a hive is healthy.

Exoskeleton

Unlike mammals, whose skeletons are located demurely within our epidermis, insects' skeletons are located on the outside. The honeybee's exoskeleton comprises three distinct parts: the *head, thorax,* and *abdomen,* each with its own highly specialized components.

Covering the entire surface of the exoskeleton are a multitude of spindly hairs. These are crucial for pollination, as pollen gathered from flowers hangs onto these hairs and holds on for the ride of a lifetime.

Head

The bee uses its head to get its bearings, orienting itself to the world around. The head is where you'll find the bee's sensory organs of taste, smell, sound, sight, and, to a lesser extent, touch.

Eyes

Honeybees have five eyes, two large ones on each side of the head, known as *compound eyes,* and three smaller eyes clustered in a triangle in the center of the head, called *ocelli.* The compound eyes are used for vision as we understand it, allowing bees to detect yellow, blue, and green shades in the color spectrum, along with ultraviolet light, which is not visible to humans. With the ocelli, bees detect light levels, allowing them to orient themselves inside the darkness of the hive.

Antennae

Sometimes referred to as *feelers,* the two antennae attached to a bee's head are its version of a nose. Capable of detecting scents at concentration levels imperceptible to humans, bees use their feelers to search out fragrant lilacs and blooming tulip poplars, nearby water sources, their hive-mates, and every other scented thing under the sun. Antennae also process touch sensations.

Mandibles

Two jawlike mandibles comprise part of a bee's mouth and are used in different ways depending on the bee's gender and caste. Tasks performed by the mandibles include molding wax into honeycomb, moving objects around the hive, gathering up pollen, and feeding larvae.

Proboscis

Used to slurp up nectar from flowering plants, the proboscis is the honeybee's version of a tongue. When a bee is not foraging, its proboscis is at rest, retracted up into the mandibles. When it's time to guzzle nectar, the proboscis unfolds into a narrow strawlike tube measuring 1/4 inch (6 mm) long.

Thorax

The "core" of a honeybee's body, the thorax is the source of all locomotion. Additionally, the thorax is crucial to gathering pollen and propolis, a resinous, glue-like substance sourced from trees and plants and used by honeybees to seal any crack or tear in the hive that needs securing or repair.

Wings

Honeybees have two sets of wings. Tiny muscles attached to the wings control minute changes of position, enabling bees to dart and change direction with rapidity and ease.

Legs

If I had a honeybee's legs, I'd be able to accomplish six different things at once. The ultimate multitaskers, three sets of legs, each with their own specific abilities, work in concert to let bees complete numerous jobs simultaneously. Interior brushes on each leg are used to remove pollen bits from the

Bee Anatomy 101

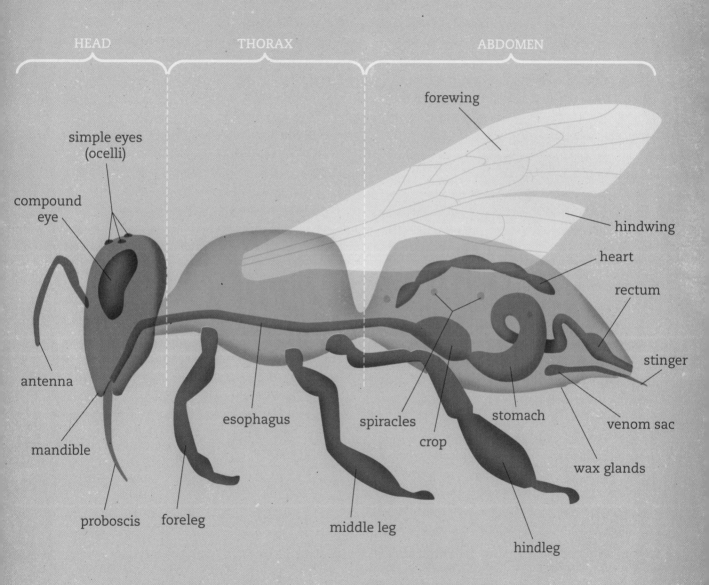

HEAD   THORAX   ABDOMEN

forewing

simple eyes
(ocelli)

compound
eye

hindwing

heart

rectum

antenna

stinger

esophagus

spiracles

stomach

venom sac

mandible

crop

wax glands

proboscis   foreleg

middle leg

hindleg

body. The front pair of legs cleans the antennae. The middle pair takes up pollen and propolis, passing it back to the hind legs. The middle legs are also used in walking, grooming the wings, and, in worker bees, grabbing onto the wax secreted from their abdomens to make honeycomb. Worker bees' hind legs contain tiny devices called *corbicula*, where pollen collected during foraging flights is stored. Attached to each leg is a tiny foot, consisting of two claws and a central pad. This claw and pad powerhouse lets bees hang onto surfaces both smooth and rough.

Abdomen

The abdomen is where all of a honeybee's most precious bits and pieces are housed. From digestive organs to those used in respiration and reproduction, the abdomen contains parts essential to both a bee's life and its potential progeny.

Respiration Organs

Running the length of a bee's abdomen and thorax are the tiniest of holes called *spiracles*. These spiracles allow the honeybee to breathe, taking in oxygen and sending it along to trachea and air sacs. Tracheal mites, a parasite we'll study in greater detail in chapter 9, enter the trachea of honeybees through the first spiracle.

Digestive and Cardiovascular Organs

A heart, true stomach, intestines, elimination organs, and a honey stomach, or *crop*, are all found in the abdomen. The digestive systems of bees are comparable to humans'; food travels down a long esophagus and is digested in the stomach. The intestines take up all of the nutrients extracted from food and then shuttle the waste along to elimination organs. A long tubular organ, the honeybee's heart runs throughout the abdomen, moving *haemolymph* (bee blood) about its entire body instead of just into individual blood vessels, as in humans.

Reproductive Organs

Honeybees, like most creatures, are sexually divided into males and females. The queen and worker bees all possess ovaries and sperm storage areas (the queen, in her role as chief baby-maker, has considerably larger versions of both). Drones contain all of the assorted and sundry "equipment" needed for fulfilling their reproductive roles, including testes, penis, and seminal vesicles.

Wax and Scent Glands

Worker bees have several additional organs, used in communication and building. Wax glands produce liquid wax, which is squeezed out from the underside of the abdomen. When the wax firms up it hardens into scales that bees chews on with their mandibles and work into honeycomb cells. Worker bees' scent glands secrete various pheromones, chemicals that convey messages to other bees.

Stinger

Pity the poor drone! He's the only honeybee lacking a stinger. Used in defense, a stinger is attached to the backmost end of the abdomen of worker and queen bees. The bee venom housed inside the stinger delivers a payload of painful sensations to the victim of a sting. Use of the weapon exacts a price, too: once a bee deploys its stinger and flies away, the stinger tears out of its backside, resulting in its ultimate demise.

17

HIVE HIERARCHY

A honeybee hive is a highly orchestrated, synergistically operating entity. Each bee's distinct role contributes not just to its own well-being but to the hive's success and continued survival. Tasks and functions are divided into "castes," or social roles, determined biologically before birth. A honeybee literally does what it was born to do. It's not necessarily all honeycomb and happiness inside the hive, however. If a member fails to live up to its role or succumbs to illness or injury, the colony may decide that its continued residency in the hive is detrimental. Then the honeybee in question is promptly (and, often, forcibly) removed or even killed. Optimal function is the hive's chief objective.

The Queen Bee

While she doesn't rule the hive, per se, the queen bee certainly regulates it. Through her single-minded devotion to her task, the queen ensures the literal survival of the species. As egg-layer-in-chief, the queen spends her days positioning her rear over honeycomb cells, depositing eggs, and doing precious little else (no wonder, when you consider the energy it would take to lay up to 2,000 eggs a day!). She leaves the hive only once in her lifetime, for an airborne amorous embrace with a gaggle of pheromone-driven drones, and then returns to live in darkness and perform her duty to the hive.

In addition to making babies, the queen's presence ensures that the hive will be on its best behavior. The pheromones she releases prompt forager bees to keep gathering nectar and pollen, guard bees to stand sentry, and nurse bees to attend to the young, among many other messages. In her absence, or when she ages and begins to fail in her duties, the hive can lose its direction. When a queen dies or begins to decline (signaled by a change in pheromones), a hive will raise a new queen. If eggs or larvae less than three days old are present, worker bees will choose one to become the new queen and begin feeding her *royal jelly*, a nutritious milky-white substance secreted by nurse bees. If no eggs or

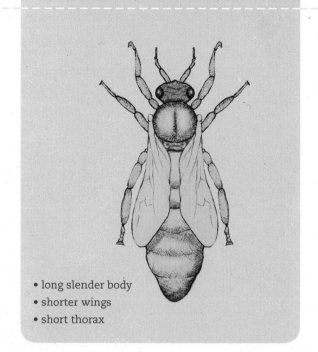

- long slender body
- shorter wings
- short thorax

young larvae are available, latent ovaries in worker bees—no longer set to "off" by the queen's scent—begin to develop. As workers are incapable of mating, the eggs they lay will be unfertilized, developing into drones. A hive populated almost exclusively by drones cannot survive for long.

The longest-lived member of a colony, a queen can live for two or more years, provided she continues to lay eggs regularly. A queen may succumb to old age or illness, or she may be a victim of *supersedure*, in which a new queen kills an existing queen. Pretty much as soon as the newly hatched queen emerges, the existing queen is thrown out of the colony, left to starve and die outside the hive, or stung to death and removed by undertaker worker bees.

When examining a hive during routine inspections, you'll always want to attempt to locate the queen. I say attempt because in some species of honeybee, her dark-colored body can be difficult to find. Look for her long, slender body, which distinguishes her from her royal subjects. The queen also has shorter wings than workers or drones, covering only about two-thirds of her body. You can also spot the queen by her short thorax, which is considerably smaller than that of worker bees.

Drones

Male honeybees, or drones, have one purpose in life: to mate with the queen. They do this by making periodic flights, searching for an on-the-prowl midair queen. Aside from that, they are cleaned up after, fed, and otherwise cared for by their busy little sisters. Males don't even take the time to do their "business" outdoors, leaving soiled honeycomb in their wake for others to clean up. I can only imagine all the eye-rolling that must go on in the hive.

With about a hundred female workers per drone, you might think the males of the hive have got it made. I beg to differ. Lacking stingers, drones cannot defend themselves, should the occasion call for it. Furthermore, they lack the ability to feed themselves or even seek out food. Toward the end of summer, when nectar stores are low and no food is being collected, drones that have failed to mate get kicked out of the hive. As if that weren't bad enough, a drone's passionate embrace with a virgin queen will also be his last. When the queen emerges from her coupling, the drone's genitals re-main attached to her, leaving him to fall from the sky and die.

Drones, like the queen, are relatively easy to spot in the hub-bub of the hive. Their bodies are longer than those of worker bees, yet smaller than the queen's. They have fat, rounded bot-toms and lack stingers, pollen baskets, or wax glands. Perhaps the drone's most distinguishing physical attribute is its large compound eyes. These eyes are much larger than those of the queen or worker and meet at the top of the drone's large head.

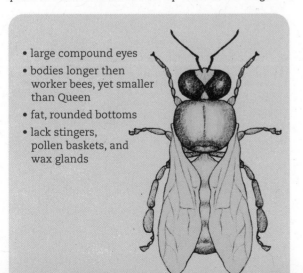

- large compound eyes
- bodies longer then worker bees, yet smaller than Queen
- fat, rounded bottoms
- lack stingers, pollen baskets, and wax glands

Worker Bees

Aside from reproductive activities, a worker bee is respon-sible for every other task required within a hive. Its roles vary throughout its short lifetime. As it ages, its sting and wax glands develop, enabling it to perform activities it was previously not physically capable of. During its lifetime, a worker bee either works from home or heads out into the world to perform its duties. These "house" or "field" activities are determined by its age. Accordingly, when you happen upon a bee out on a foraging expedition, know that you are encountering a more senior member of the hive. Worker bees born during the sum-mer months live only about six weeks, liter-ally working themselves to death. Those fortunate enough to be born late in autumn, when there's less work to do, may live up to six months, providing warmth and necessary functions for wintertime survival.

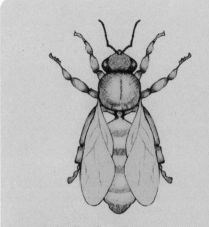

- much smaller than drone and queen
- broad food glands
- scent glands
- pollen basket

It's never a challenge to locate worker bees in a hive. Open the cover and they're pretty much the first thing you see! Much smaller than the queen or drones, worker bees possess several distinguishing specialized structures. Brood food glands, scent glands, wax glands, and pollen baskets provide all of the tools these hardworking busy bees need to make the hive thrive.

WORKER BEE DUTIES

Housework

The following in-house duties are performed during the first three weeks of a worker bee's life.

NURSING

Feeding and caring for growing larvae

ATTENDING

Grooming and feeding the queen

CLEANING (HIVE)

Cleaning honeycomb cells and keeping the hive debris-free, including removing dead bees to the front of the hive

CLEANING (BEES)

Cleaning dust, pollen, and debris off other bees

UNDERTAKING

Removing dead bees from the hive

BUILDING

Secreting wax from wax glands and rendering it into honeycomb

CAPPING

Secreting wax from wax glands and capping pupae and ripened honey cells

PACKING

Taking pollen from returning foragers and putting it in cells to eat later

RIPENING

Fanning honey in cells to remove water, thereby preserving it

REPAIRING

Placing propolis over any cracks in the hive or any foreign matter too large to carry out

Fieldwork

Once its mandibles and stinger are fully formed, a worker bee is ready to take on the outside world. At three weeks of age it heads out in the wild, fully equipped for foraging duties.

**COLLECTING
(NECTAR, POLLEN, PROPOLIS, AND WATER)**

Gathering up food and water supplies and transporting them back to the hive via either honey sacks or pollen baskets located on a bee's legs

GUARDING

Stinging interlopers and releasing an "alarm" pheromone as a warning to the hive of impending danger

BECOMING ADULTS

Before a bee becomes a bee, it must pass through several developmental stages: egg, larva, and pupa. These three stages are collectively referred to as brood. Each caste goes through a similar sequence, but important distinctions in fertilization and length between stages dictate the caste a bee will be born into.

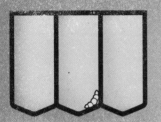

Egg

A queen is a fastidious mama. She'll only lay eggs in the most scrupulously clean cells available, prepared for her by worker bees. Once the queen finds a cell to her liking, she backs in her rear and deposits a single egg at the very back of the cell. Unfertilized eggs will become drones, while fertilized eggs become worker bees, and, on occasion (if the hive decides it needs a new one), queens. The ratio of females to males is carefully controlled in the hive by worker bees, who construct larger cells for males and smaller cells for females in very specific amounts. The queen recognizes the cell size and deposits the correct egg accordingly, usually concentrating worker eggs in the main body of the frame and drone eggs at the farther and lower ends of the frame.

Looking like the tiniest grains of white rice imaginable, the presence of eggs in honeycomb cells is a clear signal that you've got a healthy, laying queen. The eggs are incredibly small, measuring around 1.7 mm, so spotting them is no mean feat. Once you learn what to look for, locating eggs will become easier. A pair of reading glasses or magnifying goggles may help until you get more skilled at eyeballing eggs directly. Holding the frame up to the sunlight during inspection helps as well. Eggs develop for three days before moving to the larval stage.

Larva

Honeybee larvae are called grubs. At this point, they look like pearly white, glistening, curled-up semicircles. Over the next few days, the larvae are fed first royal jelly, then weaned to a blend of honey and pollen. If the hive decides to produce a new queen, she will continue to be fed royal jelly—lucky lady! The larvae grow rapidly during this period, eating like ravenous teenage boys, and shedding their skins five times. Around day six, the larvae stop eating and straighten themselves out in their cells. Ever-vigilant worker bees notice this and set about sealing them over with light brown wax, placing flat disks atop worker larvae and rounded disks atop drone larvae. Once snugly sealed up inside their temporary holding cells, the larvae spin a silken cocoon around themselves and sit tight.

Pupa

Over the next several days, the pupae undergo a profound metamorphosis and begin to take on the appearance of adult bees. Their previously white, moist skin gains color and texture, beginning with the emergence of the black compound eyes. The pupa then chews through the wax cap, emerging as an adult bee 7, 12, or 14 days later, depending on whether a queen, worker, or drone is in development. In total, it takes a queen 16 days in her journey from egg to adult, a worker bee 21 days, and a drone 24 days.

HIVE TALKIN'

While they might not dish gossip or gripe audibly about having to feed the babies and clean up the hive *yet again*, honeybees are in constant communication. Using choreographed movements and specific scents, they keep one another in the loop about everything from where a food source has been found to where the queen is in the hive and so much more.

Pheromones

Found in a number of insects as well as plants (and possibly also in humans), pheromones are composites of chemicals that send a specific message or cause a desired response in others. The pheromone system found in honeybees is one of the most complex known to exist. At least 15 different compounds interact, resulting in either releaser or primer pheromones. Releaser pheromones are short term and change the behavior of the recipient honeybee, while primer pheromones are long term and alter physiology.

Some pheromones are universal to all honeybees, while others are exclusive to each caste. Queen-specific pheromones include the all-important *queen mandibular pheromone*, which regulates hive activities, induces swarming, and even halts the ovarian development of worker bees, among other things. Drones come equipped with a pheromone that tells other drones to meet up in the sky at a certain location to mate with a queen. Worker bees, busy little creatures that they are, come with a virtual pheromone arsenal, enabling them to do everything from sending a "we're over here" aroma into the air for returning foragers (the *Nasonov* pheromone) to an odor shouting "Help! We're under attack" (the aptly titled *Alarm* pheromone). A well-prepared smoker (see page 51) and slow and easy movements help make sure your visits to the hive won't trigger the release of this pheromone.

Dances

When a foraging bee returns from a successful food-seeking expedition, does it keep the location a guarded secret, loading up on nectar and pollen to its tiny heart's desire? To the contrary. As soon as it gets back, a foraging worker bee turns into a dancing machine, conveying detailed information about a food source's location with, while not exact, a rather good degree of precision.

The round dance says "nectar nearby," while the waggle dance conveys "you've got some distance ahead of you, friends." With each dance, the in-the-know bee dances around the hive in a circle, crisscrossing and making figure-eight patterns to provide site specifics. Just how valuable the nectar source is can be gauged by the length and intensity of the dance. Discovered by German researcher Karl von Frisch, who won the Nobel Prize in 1973 for his work, a bee's dance is always based on its orientation to the sun. Accordingly, the same site discovered by a bee at a different time of day will result in a slightly different dance on account of the sun's changing position.

Honeybees engaged in communication

Profile of a Beekeeper

Jenny

Jenny's foray into the world of beekeeping was prompted by a visit to the home of her late grandfather, himself a former beekeeper in his native Czech Republic. Once she returned stateside, Jenny signed up for a local course on the subject, but it would be another seven years before she would take the step toward ordering her first package. Now the proud steward of several thriving hives, this bicycle mechanic and educator, steel sculptor/welder, and former glassblower feels connected to her grandfather again, often turning to him in her prayers for beekeeping guidance and wisdom. Her hobby has also caused a swell of pride in her mother who "marvels at the continuity of our family's love of bees." In addition to her family's beekeeping legacy, Jenny details her primary reason for keeping bees is just "the sheer joy of it, with all the wonderful by-products, experiences, connectedness, and gardening benefits being supporting reasons."

While her beekeeping experiences have been characterized largely by enjoyable experiences, Jenny does admit freely to a good bit of initial apprehension. "There was uncertainty, fear, anxiety, and all ranges of emotional stress, all stemming from the fact that I had about ten thousand new charges in my care! I didn't want to do anything wrong, didn't want to hurt a single bee, but there's no avoiding the occasional mistake, misstep, or sad sound of crushing a small life as a new beekeeper." Moving into her third year of beekeeping, Jenny still frets about her bees in bad weather, but rests confident in the knowledge that, as she puts it, "as long as you have love in your heart, it will see you through the mishaps to learn and become better at every step, and that there are usually plenty of people with the same love who want to help you." I'm sure her grandfather would have been proud.

Chapter 2

What to Consider:
To Bee or Not to Bee

The siren song of beekeeping can be difficult to resist. Honeybees are fascinating to observe, essential to the pollination of over one-third of foods enjoyed by humans, and produce a number of desirable items such as honey and beeswax. Although the craft is not necessarily difficult, not everyone is a good candidate for honeybee caretaking. Like any form of stewardship, beekeeping requires investments of space and time as well as money. You're reading this book, so your interest is clearly piqued. Let's examine some preliminary considerations before you get your heart set on an adorable Italian or intrepid Russian (honeybee, that is!).

ON LOCATION

Compared to other forms of animal husbandry (dairy cows come to mind), honeybees require very little space—just what is needed for a hive and a little room to work. You can keep honeybees everywhere from urban rooftops and quarter-acre backyard plots to vast, rural expanses. Honeybees are skilled foragers and will travel miles from the hive searching for nectar and pollen.

Urban bees seem to do quite well in cities where a fairly good degree of biodiversity persists. Public parks, landscaped office buildings, antiquated churches planted long ago with ambrosial roses, rooftop gardens, backyard veggie patches, potted herbs on balconies—the amount of urban flora and fauna is profuse. Packed into such dense spaces, nectar and pollen available from these sources, along with water from lakes, creeks, rivers, birdbaths, and puddles, provide ideal foraging grounds for our buzzing friends.

Honeybees are capable of surviving and thriving in a wide range of climates and terrains, albeit with a little assistance from you. You can find honeybees in rugged hillsides and flat valleys, from locales where winter means pulling out the snow shovel and layering on wool sweaters to those where flip-flops and a beach towel are all that's needed. Provided there are plants that bloom and flower, honeybees will hang their hats wherever you hang yours.

As you are considering where to place a beehive or two, keep in mind that while bees don't take up much room, the area where you locate their hives will need to meet a few key criteria. It will require some sort of windbreak, a bit of morning sunlight, and some degree of shade. You will also need access to water and room for you to navigate around the hive for examinations and extraction. We'll discuss these needs in greater detail in chapter 3 when we examine hive components and where best to situate your hive.

MONEY MATTERS

Unless your preferred pastime is cloud watching or counting raindrops, the hobbies you already engage in most likely incur a bit of expense, either at the outset or over time. So, too, with beekeeping. When you factor in the cost of the hive itself, protective gear, equipment such as a smoker (used for calming the bees), and the bees themselves, in addition to food (should you need to supply it), medications (should you opt to use them), resource materials, and vessels for holding honey, the initial expense of beekeeping can seem daunting. While I will admit that the startup costs are considerable, there are a number of ways to begin beekeeping without breaking the bank.

A local beekeeping chapter is a great resource. Such organizations may have periodic giveaways or be recipients of donations from larger organizations looking to fund novice beekeepers. When I first began my foray into the world of *Apis mellifera*, I attended a two-weekend-long beginning beekeeping

An assortment of beekeeping gear

school sponsored by my regional beekeeping organization. All attendees were entered into a contest for a litany of apiary goods to be given away at the conclusion of the school. I won an expertly written book on beekeeping, while other attendees went home with everything from smokers to full hives. Many chapters have more costly equipment, such as honey extractors, available for rent. My local chapter rents out its mechanized extractor for a mere $11—the same equipment costs hundreds to purchase.

It might be tempting to purchase secondhand beekeeping equipment or hives. While something like gently used protective clothing and gear poses no threat, used hives might harbor disease. This is where your local chapter can assist you in connecting with a reputable seller in your area. You could also contact the regional governmental representative for beekeeping concerns, such as a state or county bee inspector, to solicit advice on used bee goods. If you elect to buy new, I'd encourage you to shop around. When first buying my hives and equipment, I found a great variety in prices between suppliers. Search for deals, but don't scrimp too much. When it comes to beekeeping equipment, it pays to start with the best. Going cheap all around will only require greater investment later when bargain gear begins to break down or fall apart.

After you've got the major infrastructure in place, the ongoing expenses of beekeeping are relatively small. You'll need sugar for feeding them when nectar isn't present for foraging; additional supers for when the honey starts coming in; new queens when the occasion merits it; jars and labels for the honey; and additional hives, if you decide to acquire more bees or split an existing hive (more on this in chapter 6). If you live in bear country, add bear fencing to the list. Overall, maintaining a hive of bees is quite affordable, especially when compared to the costs of food and veterinary care for pets such as cats and dogs.

KEEPING TIME

If you take a dog for a daily walk or clean a cat's litter box regularly, you're already devoting more time for animal maintenance than will ever be required in beekeeping. You'll spend the largest amount of time with your bees during the first year, inspecting them, looking for problems, checking that a queen is present, feeding them if necessary (we'll discuss that in greater detail in chapter 4), and other getting-to-know-you tasks. Thereafter, visits to the hive may be few.

Over the winter, you'll visit your hive only occasionally, just to check in and make sure nothing has caused any structural damage (like a predator, strong gust of wind, or fallen tree branch). As the warmer months approach, it will be necessary to stop by every week or two to check the hive's progress. When honey is available, several hours will be required for removing the supers, extracting the liquid, and cleaning the equipment when done. A bit of fall maintenance to prep your hive for the impending cold weather closes out a year's activities for the beekeeper. I'll offer more information about the best time of year to get started with bees, as well as seasonal upkeep, in chapters 6 and 8.

While there isn't a great deal of daily care involved in beekeeping at any given time, bear in mind that your bees will need to be looked after regularly. If you're the type to take lengthy vacations during the summer months, or get pulled away at work for weeks at a stretch, it would behoove you to find a beekeeping friend before setting up your hives. A reciprocal arrangement with a fellow beekeeper, whereby you offer to care for one another's hives when life calls you away from home, can be especially advantageous. A "bee-sitter" can drop in, inspect the hive, provide food if needed, and add a super or two if honey is beginning to fill up.

LAW-ABIDING CITIZEN

While many places in the world gladly welcome beekeeping, there are those that do not. In the United States, each state has its own rules on beekeeping. Within those states, variations in what is and isn't allowed exist from one municipality to the next. Before you place an order for bees, gear, or equipment, bear in mind that some cities forbid beekeeping outright, while others place restrictions on it, such as hive location or number of hives per residence. Some renegade apiarists house bees in cities that prohibit doing so while simultaneously trying to have the laws changed in their favor. These "outlaw" beekeepers accept the risks that such a practice may incur (fines, relocation of hives, and so on). Check the codes in your city—available through the Department of Agriculture, local zoning board, town hall, bee club, or governing body—to determine what is allowed.

BEING NEIGHBORLY

The popular saying "good fences make good neighbors" is especially apropos when it comes to beekeeping. If your neighbors are within your direct line of sight, I would urge you to mention your beekeeping plans beforehand. Unfounded or not, some people have fears about bees. Popular ignorance and misinformation about bees has resulted in widespread opinions that are simply not true. It's your job, then, to raise awareness, dispelling myths before they fester into false beliefs. The following tips will go far toward assuaging your neighbors' concerns about your winged, foraging friends.

→ Start with gentle bees to begin with. If you find your bees to be aggressive, consider requeening. This often calms the hive within a month or so.

→ Be inconspicuous. Place your hive in a location that doesn't put neighbors, pets, or children in a direct flight path. Returning foragers follow a routine pattern and could potentially sting anything that crosses their path. If your neighbors are quite close, plant a high hedge or install a fence at least 6 feet (1.8 m) tall to encourage bees to fly up, over, and away from anything nearby. Also, position the hives to face away from children's playgrounds, dog kennels, and neighbors' doors.

→ Many complaints from neighbors are based on thirsty bees drinking from their birdbath, swimming pool, pond, or other water source. Once bees discover a water source, it's difficult to get them out of the habit of visiting, so provide water near the hive and put a stop to this situation before it begins. I'll cover water options in greater detail in chapter 3.

→ Don't keep more than four hives on property smaller than a quarter acre. When you are working the bees, or if a colony should swarm (leave the hive in search of a new home), all those buzzing bees can make the neighbors nervous, even though you might know there's nothing to be worried about (a swarm is actually quite docile; they're just looking for a place to start setting up house and care for the babies).

→ Remember, as the saying goes, you can catch more flies with honey than with vinegar. Ply your neighbors with honey. Bake them sweet delicacies made with your honey or craft them some homegrown beeswax candles. Invite them over to watch as you examine the hives, showing them just how good-natured and rather indifferent to your actions the bees are. Let them know that the life of a beekeeper, their friends, and neighbors is one characterized by sweetness, in word, deed, and flavor.

Avoiding Bee Stings

While the occasional sting may happen, there are a number of precautionary steps that can be taken to minimize the risk.

→ Wear protective gear when working with your bees, including gloves, a veil, and thick, light-colored outerwear. Tuck pant legs into boots or shoes to prevent bees from sneaking in—they love to explore dark cracks and crevices.

→ Move slowly when working the bees. Quick, jerky movements make them concerned that danger is approaching.

→ Smoke the hive entrance before opening it. This will help calm the bees. Be sure your smoker is well lit before you begin your examination and that it will remain lit for the duration of your visit (I'll discuss lighting a smoker on page 51-52).

→ Situate your hive so that the entrance doesn't directly face children's play areas or other heavily trafficked areas. Keep in mind, though, that the best hive positioning is to face east or southeast to take advantage of morning light. Similarly, don't block the hive entrance when working with the bees. Stand to one side or at the back.

→ Avoid working your bees on inclement weather days. Also, try to avoid working the bees either too early or too late in the day. During bad weather, early mornings, or twilight hours, most of the hive will be at home. It's best to visit when a good number of the hive's residents are out foraging.

→ Try not to bang, hit, or otherwise jar the hive. Strong vibrations upset the bees. Be gentle when working the bees and they will react in kind—in fact, they'll most likely just ignore you.

PETS & CHILDREN

Pets and children share one defining trait: curiosity. When it comes to beekeeping, this is often good for children, while not so ideal for pets.

When introducing children to bees, one hopes to kindle a natural curiosity about the hive, its occupants, and its goings-on. I've seen kids as young as five who are active beekeepers, assisting Mom or Dad with hive chores, honey extractions, and the like. The lessons on pollination, synergistic workload, and hive interdependence are fantastic learning opportunities for children. Beekeeping gets the little ones outdoors and moving (albeit slowly and gently, when actively working the hives), breathing fresh air into their developing lungs, and experiencing the living world around them. Bee suppliers carry protective gear in children's sizes, and a number of books and videos written expressly for educating children about honeybees are available. If you encounter a child who harbors a fear of bee stings, do your best to explain to them all of the precautionary measures that can be taken to prevent stings from occurring, as well as the fact that honeybees really don't want to sting humans. You could also put together a "just-in-case" sting kit, with ointment and colorful bandages. Show it to your young beekeeper to assure them that help is available should they ever need it, and put it in a readily accessible location, easily reached by hands large and small alike.

As for pets, many will quickly learn that the hive is nice from a distance, but something else entirely if they move in too close. Cats and dogs might take a sniff or two, or even move toward the front of the hive, but most will back away as soon as they get a sense of what's in there. My dogs and cats have always steered clear of the hives, perhaps knowing on some level that the bees are not their friends. While they are out foraging, honeybees are of no real threat to animals. They are extremely focused when foraging, and an attack is highly unlikely. That said, if your pets exhibit too much curiosity, a barrier of some sort might be in order. Larger animals such as goats, sheep, or horses should definitely be prevented from accessing the bee area, as their movements could knock over the hive.

As I've had a close encounter with a bear on my property (see page 93), I keep my hives surrounded by electric fencing. Once, while I was outside in the chicken coop, my dog must have ventured a bit too near the fence while sniffing out some creature or other. I heard a sudden yelp, and then saw him running from the direction of the hives. He continued looking in that direction for the next few minutes with a "there's something mean over there!" look on his face. To date, he's shown no interest whatsoever in exploring the hives.

ALLERGIES

Honeybees take all the blame. Fingers get pointed at *Apis mellifera* for stings that actually came from wasps, yellow jackets, hornets, bumblebees, carpenter bees, sweat bees, and many other insects. Any sting will result in swelling and redness at the site of the wound. The subsequent pain and soreness will last anywhere from a few hours to several days. This is entirely normal.

However, for those with an allergy to bee venom, a bee sting may cause life-threatening anaphylactic shock. While this condition occurs in less than one percent of the general population, it is extremely dangerous. An allergist can conclusively diagnose allergies to honeybee venom. If you or a member of your family are allergic to bee venom, consult with your physician for the best course of action. They may suggest carrying an epinephrine auto-injector, in which sensitive individuals self-administer an antidote to bee venom. Conversely, you may choose to abstain from beekeeping altogether. There are plenty of ways to enjoy and assist honeybees without necessarily keeping them yourself. You could plant a garden with our many pollinators in mind (in an area not directly near your home, such as a community garden or church yard), attend a lecture at a bee club, or participate during the extraction of a beekeeping friend's honey.

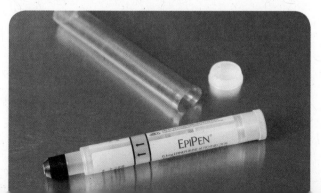

Profile of a Beekeeper

Megan

When she's not busy working with a Manhattan children's clothing designer, Megan is busy working her honeybees. A rooftop beekeeper from her decidedly urban outpost in Brooklyn, NY, she tends to two Langstroth hives and plans to soon try her hand at a top bar hive, increasing her apiary size to three. The operator of Brooklyn Honey, Megan sells the honey and comb her bees gather from around the Five Boroughs, including nectar sourced from plants in Central Park. Beekeeping satisfies Megan on so many levels. "I have always been interested in gardening, insects, eating joyfully, being outside. Beekeeping touches on all of those, so it's been a really good fit for me so far."

Childhood summers spent on her family's Virginia farm introduced Megan to honeybees via beekeeping neighbors. Now a city dweller, beekeeping creates a bridge between her current stomping grounds and the greater, greener outdoors she craves. "The agrarian lifestyle definitely speaks to a part of me, and honeybees are the easiest way for a city dweller like myself to forge that connection to the land."

Megan has chosen Italian honeybees to occupy her hives for a number of reasons, claiming "they were recommended to me for their manageability, propolizing habits, docile temperament, and lesser propensity for swarming." This urban beekeeper, who also keeps chickens and grows vegetables and herbs in her backyard plot, recommends keeping an open mind and a curious spirit when entering into the world of *Apis mellifera*. As she puts it, "No one book will have all the answers, nor will any single beekeeper have all the right advice. Cast your net wide and find the advice and techniques that work best for *your* needs, *your* bees, and in *your* region."

SPECIES VARIATIONS

Next you will need to consider what species of honeybee to acquire. Having evolved in different geographic regions, different honeybee species possess different traits and characteristics. A bee hailing from a cold region, such as Russia, might manage harsh winters better than its sun-loving Italian cousin, for example. In addition to hardiness, bee species vary in terms of their aggression, resistance to pests and diseases, and their propensity for *robbing*. Robbing is a truly undesirable situation, when a stronger hive (or population of wild pollinators, such as wasps) invades a weaker hive, stealing honey. I'll talk more about preventing robbing in chapter 9.

Honeybees belong to the genus Apis, meaning "bee-like" in Latin, of which there are eight species. *Apis mellifera*, literally "honey-bearing bee," is one of these species and is further broken down into 24 races, or subspecies. The following are some of the most popular and common races of honeybees kept by beekeepers, which are readily available from bee suppliers.

Carniolan (A. m. carnica)

Hailing originally from east-central Europe, the Carniolan was introduced to North America in 1883 and can now be found across the world. This variety of honeybee is dark-colored, with gray bands running across its body instead of yellow. Advantages of Carniolans include an exceptionally gentle temperament, good management of winter food stores, rapid build-up of brood in the spring, small robbing tendency, and a general likelihood toward surviving winter conditions (see page 82).

Caucasian (A. m. caucasica)

This race of bees is named after the Caucasus region, sandwiched between the Black and Caspian seas. These bees are visually characterized by their gray-black bodies. Advantages include a very gentle temperament, the ability to overwinter well, and the longest proboscis of all European honeybee races, which allows them to extract nectar from a wide range of plants. Disadvantages include a tendency toward swarming, liberal use of propolis (which can make examining the hive challenging to the beekeeper), and susceptibility to certain diseases such as nosema (see page 86).

German (A. m. mellifera)

German honeybees, also known as "black" or "north European" bees, were the first species imported by early American colonists to pollinate European varieties of fruits. Once the most common honeybee in North America, German bees dominated the landscape of their new environs until the introduction of Italian bees. Although hardy producers of brood and honey, Germans had a reputation for aggression, prompting American apiarists to search out a more docile bee.

Italian (A. m. ligustica)

Discovered in 1859, the gentle Italians are now one of the most popular species of honeybees kept in North America. Characterized by yellow-golden and brown bands, these honey workhorses hail from the Apennine Peninsula in the "boot" of Italy. This species all but completely replaced the German bee when introduced to the United States, as its relative lack of aggression, coupled with intense honey production, made it a fast favorite. Italians are readily available from most bee suppliers. Advantages of this species include production of a large amount of brood and honey, buildup of large winter colonies, and low swarming tendency. Disadvantages include robbing tendency, as large winter colonies often eat through their food stores too soon, and susceptibility to certain pests and diseases (although not European Foulbrood, to which they exhibit marked resistance; see page 85 for a detailed discussion).

Russian

The U.S. Department of Agriculture, in cooperation with the Russian government, began a research project in the 1990s, looking for a species of honeybee with resistance to varroa and tracheal mites, two parasites with a reputation for incurring large losses of honeybee populations (see pages 87-89). After discovering that Russian bees had long lived with and developed resistance to the haemolymph-sucking pests, importation began in earnest. A hardy stock, this variety of bee has faired rather well in North America. Russians resemble Carniolans in a number of ways, including slowing down the buildup of brood (i.e., baby bees) and food consumption when nectar supplies are low, leading to smaller winter colonies. They are comparable to Italians in a heavy production of brood and honey when nectar and pollen are readily available. Russians are now available from many bee suppliers.

Buckfast (hybridized)

Developed by Brother Adam, an English Benedictine monk, in the 1920s, Buckfast bees are named after the abbey in which they were first produced. By carefully cross-breeding several races, Brother Adam created a docile, highly productive bee that exhibited strong resistance to tracheal mites and overwintered well. One disadvantage of Buckfast colonies is the need to requeen annually. As members of a hybrid species, Buckfast offspring queens will not "breed true," with traits identical to their mother. Therefore, supercedure, if allowed, would result in genetic drift within the hive. Buckfast bees also have a tendency toward slow buildup of brood in the spring and are therefore not the best choice if you're looking to pollinate early spring crops (Carniolans are better suited for that purpose).

Starline (hybridized)

The world's only hybrid race of Italians, Starline honeybees were developed to produce well when used for commercial purposes. This strain of bee builds up a large amount of springtime brood, produces ample honey under the right conditions, and uses a minimal amount of propolis, allowing for easy inspections and frame removal come extraction time. On the down side, Starlines may have difficulty overwintering; as with most Italian honeybees, their large colonies can create wintertime food scarcities. Like Buckfast bees, colonies may need to be requeened annually to prevent genetic drift.

Africanized Honeybees (AHB)

As a beekeeper, I can think of few things as ominous sounding as a "killer bee." Given that sensationalistic moniker by the media, these bees are known to apiarists as Africanized honeybees. In the late 1950s, a group of Brazilian researchers were attempting to create a hybridized bee capable of producing plentiful honey while withstanding tropical heat and humidity. Such a bee could aid in pollination, offer export products, and provide work for the nation's poor. The notoriously aggressive African bee (already adapted for a warm climate) was imported and bred with the considerably more laid-back European honeybee. Unfortunately, in the process, 26 African queens escaped, fled into the wild jungles of Brazil, and bred with the wild European bee populations.

What resulted is a fiery-tempered honeybee, not to mention a public relations headache for the cause of AHB's quieter, more docile kin. *Apis mellifera scutellata*, as Africanized honeybees are technically known, possess a reputation for being intensely protective of their hives. They are quick to attack, will travel far distances to pursue an "assailant," and stay annoyed for days afterward. Africanized honeybees are, understandably, quite difficult to keep. Under intense management, however, the bees are quite productive, and Brazil's beekeeping industry has thrived. Ongoing efforts there at re-domestication of the bees over the past few decades have resulted in the development of more docile stock, which in turn can serve as an entry point for breeding future generations of gentler AHBs.

The danger, however, has arisen from swarming AHBs that have moved across South and Central America, into Mexico and, now, into a number of southern U.S. states, stretching in a band from California to Florida. Improperly managed or wild Africanized honeybees can be lethal to animals. Human deaths are extremely rare and have involved almost exclusively elderly persons unable to protect themselves from an attack. We will discuss ways to deal with Africanized honeybees in chapter 9.

Chapter 3

Housing:
Understanding Bee Space

A crucial component in keeping bees, housing is what separates the practice of beekeeping from wild colonies of bees. Here we'll examine the parts of the hive, details about purchasing housing equipment, and where to place your bees. After reading this chapter and acquainting yourself with bee housing essentials, you should feel considerably more informed about choosing proper lodging for your new companions.

ANATOMY OF A HIVE

Although composed of individual members (queen, workers, drones), a beehive can be thought of as a distinct, fully integrated, functioning organism. All of the residents' actions work toward the optimal well-being and survival of the entire hive. Should one or more members begin to fail in their duties, the entire hive is put at risk. Bees are inherently social creatures, rubbing elbows with their hive co-members not out of politeness, but out of necessity. They *need* each other, in the truest sense of the word.

Once they have found a suitable home, honeybees go about the business of forming vertical sheets of hexagonally shaped comb, produced from wax excreted from their bodies. The sheets of comb are suspended from the roof or top of whatever vessel they've elected to set up shop in. No matter whether that space is a tree trunk or a human-made hive, honeybees maintain a minimum distance between combs of never less than ¼ inch (6 mm) nor greater than ⅜ inch (9.5 mm). These regular intervals allow them to easily navigate around straight, evenly spaced combs. That space, with its rigidly determined parameters, is referred to as *bee space*. If the bees encounter a space larger than bee space, they build additional comb. Too-small spaces will be filled with propolis so that no smaller interloper can wiggle its way into the hive.

HIVE COMPONENTS

While several different models of beehives are found around the world, the most commonly used form is the Langstroth hive. As described on page 13, Langstroth invented the moveable hive in the mid-19th century. Prior to his invention, most hives needed to be seriously damaged or destroyed for honey extraction and beekeeping inspections. The Langstroth hive is essentially a series of stacked wooden boxes. These boxes build on the bees' natural style of honeycomb building, permitting the hive to build following the principles of bee space mentioned above.

Through attentive observation, Langstroth was able to determine the exact parameters guiding honeycomb construction. He then created housing appropriate to the honeybees' unforgiving specifications. Once comb is built, bees fill it up with brood or food, and then move on to the next available space. Adding boxes, or *supers*, on top of existing boxes permits the bees to increase both the hive's size and food supply. The stacked boxes are filled with removable vertical frames. These frames resemble flat drawers and are the surface upon which the honeybees build their comb. Beekeepers can easily inspect each frame for hive health and remove them as needed for honey and wax collection.

The components of the Langstroth hive are known as *woodenware*, as they are usually made of pine, cedar, or cypress, although a number of plastic models are also becoming available. We'll examine the function of each component below and on the following pages.

Hive Stand

The hive stand is the primary support for the entire hive. Keeping the hive off the ground is quite important, for a number of reasons. First, excess moisture can be a death knell for bees. Raising the hive off the ground prevents precipitation from pooling or condensing in the hive. Second, elevating the hive moves it out of easy access from most other insects and opportunistic predators, as well as preventing low-growing plants from blocking the entrance. Lastly, having the hive above ground level keeps the beekeeper from needing to bend and stoop over to examine the hive, feed bees, remove frames, or do any other hive-related activity.

Hive stand

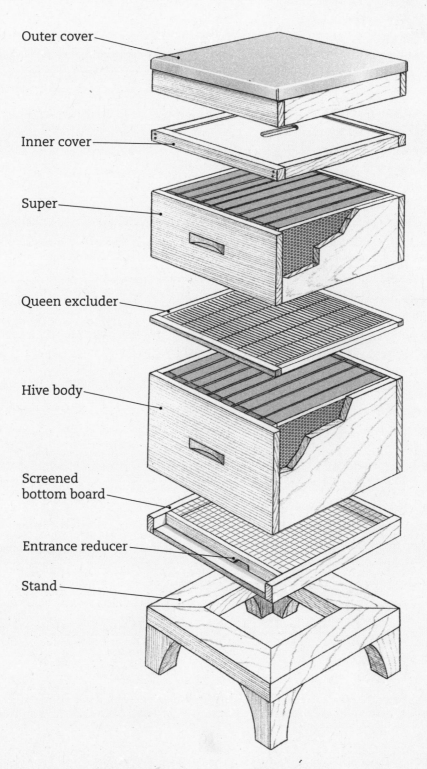

Outer cover

Inner cover

Super

Queen excluder

Hive body

Screened
bottom board

Entrance reducer

Stand

While most hives come with a wooden stand consisting of three side pieces and a gently sloped landing board, many beekeepers feel the hives are not quite high enough using only those pieces of woodenware. Most prefer to raise them somewhere between 2 to 3 feet (61 to 91 cm) off the ground. This can easily be achieved either by constructing a wooden stand out of 2x4s, or resting the hive atop concrete blocks, railroad ties, or some other strong support that can withstand up to 200 pounds (91 kg) of weight (a possibility with an active colony in full honey and brood production). If you opt to use concrete blocks, as I do, you might want to place a flat board of heavy lumber between the blocks and the underside of the hive, to give you a workspace to rest tools or hive components during examinations.

Bottom Board

Essentially the hive's floor, the bottom board also contains the entrance to the hive. Bottom boards may be made from solid pieces of wood or from metal screening. Screened bottom boards are beneficial for providing ventilation and temperature regulation to the hive during warm summer months. They can also be used to monitor and control the population of the parasitic varroa mite. When mites are a problem, beekeepers can either allow mites to simply fall out of the hive (preventing them from crawling back in) or place a sticky mat atop the screen, which traps falling mites, and count the population to determine the rate of infestation (more on this in chapter 9).

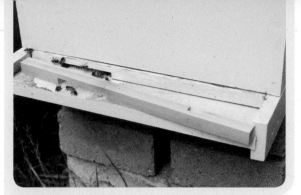

Entrance reducer

Screened bottom board

Entrance Reducer

This small, removable piece of woodenware is a strip of wood running the length of the hive's entrance with carved notches of varying size on both sides. The different notches provide openings of varying size, permitting the beekeeper to regulate the degree of access to the entrance. Under ideal circumstances during warmer months, the larger entrance is used. When the weather cools, the smaller side is installed, in order to prevent chilly mice looking for a free place to couch surf from entering the hive. The smaller entrance may also be utilized if robbing should ever occur. We'll talk more about preventing robbing in chapter 9, but know for now that an entrance reducer can be a stopgap measure for dealing with such a scenario.

Hive Body

Also known as *brood chambers, brood boxes, deep hive bodies,* or simply *deeps,* hive bodies are the largest boxes in the hive. They sit directly atop the bottom board and house the queen, brood, and the food the bees eat themselves. Hive bodies are about 9 ½ inches (24.1 cm) tall and can hold 9 to 10 frames (more on frames on page 36). Many are made of wood, either pine or cypress, although there are plastic models available. These structures have four sides but no top or bottom, so that bees can climb from one box to the next inside the hive. They have recesses cut into their sides to allow for lifting or hoisting, should the need arise.

The decision to use one or two hive bodies is usually determined by the climate in your area. If you live in a region where harsh, extended, truly cold winters do not occur, then you should be fine with one hive body. Those living in cold-weather areas may need to use two hive bodies to provide enough space for brood and food to last all winter long. A combination of one hive body and one medium super or a collection of three medium supers will achieve the same goal. When full of honey, wax, bees, brood, nectar, and pollen, a hive body can easily weigh between 60 and 80 pounds (27 and 36 kg), a hefty burden for even the most muscular among us. Some beekeepers use a configuration of smaller supers to create both brood chambers and honey storage. Using a consistently sized super throughout your hive allows for the interchange of frames, should you need to move things around.

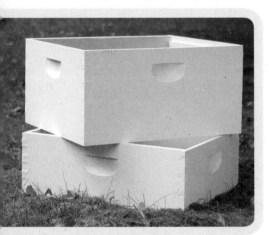

Supers come in varying sizes.

high and will weigh somewhere between 40 and 45 pounds (18 and 20 kg) when full. As a petite woman, I've found a combination of one deep hive body, topped with a medium super, followed by additional shallow supers, to serve me well. Select the setup that best suits your needs based on your physicality and activity level with the bees in terms of frequency of visits to the hives and opportunities for extraction (mediums will hold more honey, but will weigh more than shallows).

Frames

Frames and Foundation

Frames are the supports on which bees build their comb, "framing" out their environs. In the wild, comb would simply be built (referred to in beekeeping jargon as *drawn out*) from the top of whatever enclosure the bees choose to set up shop in. In a hive tended by a beekeeper, frames, enclosed on four sides by wooden or plastic panels, allow for easy inspection without breaking or otherwise harming the comb. Most supers hold 10 frames, although some will accommodate more. You can opt to either purchase frames fully assembled or unassembled. The cost will be somewhat less for unassembled frames, but, accordingly, time will have to be spent putting them together yourself.

Beeswax foundation

Supers

Shallow and medium supers are the boxes placed above the hive body and are used for honey storage. The honey stored in these boxes is for the beekeeper's use, while the honey and food for the bees themselves will be in lower boxes, closer to the brood. If you opt to forgo use of deep hive boxes completely, medium supers will then constitute the make-up of the hive itself, with three medium supers for housing topped by any number of shallow supers.

Medium honey supers are 6 5/8 inches (16.8 cm) high. They are also occasionally referred to as *Dadant*, *Western*, or *Illinois* supers and weigh around 50 pounds (23 kg) when full. Shallow supers are 5 1/16 to 5 3/8 inches (12.9 to 13.7 cm)

Plastic foundation

This isn't necessarily difficult, but if you're short on time, or not particularly adept at assembling things, I'd spring for the assembled frames.

Frames are usually a standard length of 17 ⅝ inches (44.8 cm). Depending on the depth of the super they are being placed in, either 9 ⅛-, 7 ¼-, 6 ¼-, or 5 ⅜-inch-deep (23.2, 18.4, 15.9, or 13.7 cm) frames are used. Standard frames consist of a top bar, two side bars, and a bottom bar. They hang inside supers on ledges called rabbets. Attached to the center of each frame is a thin sheet of either beeswax or plastic known as *foundation*. The frames provide support for the foundation, giving it structure and rigidity so it will hold up over time, particularly when subjected to the fast speeds used in centrifuges during honey extraction (see page 106 for more on extraction). Most frames have a removable top bar, so that foundation can be replaced or added as necessary.

Foundation, whether plastic or beeswax, is embossed with hexagons. The honeybees use these shapes as a guideline for drawing out the comb, producing wax cells from their bodies, and then chewing and manipulating it in place. I find the entire process stunningly beautiful, a mystical honey-bee alchemy. Whether you choose to use beeswax or plastic foundation is a matter of personal preference, as each has its advantages and disadvantages. Plastic will last considerably longer, and it resists wax moths particularly well. However, bees are somewhat slower to accept and begin building on plastic foundation. Realizing this, some manufacturers have begun coating plastic foundation with a very thin sheet of natural beeswax. Natural beeswax foundation, while readily accepted by the bees, is considerably more fragile than plastic. It requires wire supports, inserted between two beeswax sheets, in order to provide it with body and structure.

Queen Excluder

This piece of woodenware is used when honey production is in full swing. It consists of a wooden frame that encloses either a metal or plastic slatted grid. As the name suggests, it excludes the queen from certain parts of the hive. If you intend to harvest honey, a queen excluder will keep the queen from entering the supers where honey is being

Queen excluder

produced and stored. Placed on top of the hive bodies, where brood is stored, the queen excluder allows worker bees to easily pass through, but prohibits the larger queen's access. Using this tool helps to keep the developing baby bees separate from the honey. Without it, you could have a great deal of brood mixed in with the honey come extraction time. Since keeping the colony growing and thriving is the beekeeper's goal, killing off brood is to be avoided. Once the honey flow comes to an end, the queen excluder should be removed.

The use of queen excluders is somewhat controversial among beekeepers. Some argue that its use actually hinders and slows honey production by limiting the bees' movement. Honeybees are highly efficient and may be reluctant to move up into higher supers to store nectar if the brood chamber still has cells available for filling. To encourage their movement, then, it is suggested that the bees be allowed to move freely between the hive body and the supers before the queen excluder is added. This way the bees will have begun to store nectar in the supers and, again guided by efficiency, they will be encouraged to move upward and cap it for honey production.

Inner cover

Inner Cover

If you think of the hive as a house, the inner cover is the interior ceiling. Resting atop the uppermost super and underneath the outer cover, the inner cover prevents the bees from gluing the hive shut with propolis and wax. It also affords an additional layer of

Profile of a Beekeeper

Corky

Corky loves construction. Both in his day job as a remodeling contractor in the Seattle area and in his side job with Ballard Bee Company, he's continually repairing, restoring, establishing, bolstering, and fortifying human and honeybee homes alike. Ballard Bee Company is a full-service beekeeping business, allowing customers to rent one of Corky's hives, sponsor a hive in their own gardens, or have one-on-one consultations about beekeeping. For beginners, Ballard Bee will erect the hive, fill it with bees, and leave the owner to take over from there.

Corky's interest in keeping bees began in junior college where he would lend a hand to a beekeeping friend, learning as much as he could each visit. Keeping hives of his own wouldn't come until years later, when concern over dwindling U.S. honeybee populations (down from 5.5 million hives in 1945 to 2.5 million in 2005), especially in urban environments, prompted him to take up the smoker. "By having hives I felt that I could directly be involved with a solution to a serious problem." Based in Ballard, Washington, an urban community just outside of Seattle, Corky now tends more than 50 hives. While most of the hives are occupied by Italians and Carniolans, he does have several hives with hybrid queens. His bees are available to rent, own, or simply enjoy honey from. With so many hives to tend to, this seasoned beekeeper knows of what he speaks when he advises that proper care of honeybees begins with a peaceful disposition. "The best tool you can bring out to a bee yard is your state of mind. Calm mind equals calm bees."

insulation from the harsh heat of direct sun in summer and the chill of cold, moisture-laden air in winter. Inner covers have an oblong hole cut out of the center to allow air to circulate, increasing ventilation at all times of the year.

Inner covers are made of wood, Masonite, or plastic, and are flat on one side with a raised rim on the other side. When positioning the inner cover over the top of the upper super, the flat side faces down. You'll know it's on the right way if the rimmed, traylike side is oriented skyward. Some inner covers also have a small notch carved out of one of the shorter ends. This is for additional ventilation and should be positioned at the front of the hive. Inner covers are known as *crown boards* in some parts of the world.

Outer Cover

The uppermost component of the hive, the outer cover, serves as its roof. Framed by wood, the cover itself is most often made of galvanized tin or aluminum. Its sides extend beyond those of the boxes beneath it, much like a roof extends beyond the walls of a house. This allows precipitation to drip off the sides and onto the ground, keeping the hive itself dry. The outer cover fits perfectly over the inner cover, enclosing both it and the top portion of the highest super.

While the Langstroth hive is the most commonly used model around the world, it is not the only hive type in use.

NATIONAL HIVE

The most well-known hive type in the United Kingdom, the National hive makes use of smaller frames than those found in a Langstroth hive. The top bar frames in a National hive are also longer than those in a Langstroth, allowing it to hold 11 frames. The bee space, the area in which honeybees move freely, is located at the bottom of a super in a National hive, while it can be found at the top of Langstroth hive supers.

COMMERCIAL HIVE

As its name suggests, the Commercial is the hive model most favored by commercial production beekeepers. These hives have large brood boxes and are often used interchangeably with National hive supers. Commercial beekeepers like this model because it permits a large colony of bees to live comfortably in the hive, minimizing the risk of swarming.

WBC HIVE

This attractive structure was invented by Englishman William Broughton Carr, and is the more ornamental of the U.K.-designed hives. The WBC is a double-walled hive, as compared to single-walled models such as the Langstroth or National. Such hives are covered by an exterior layer of wood, with the supers and hive body nestled inside. The WBC has sloped sides and a gabled roof and must have the outer panels removed before the interior may be accessed. While gaining entry to the hive may be more challenging for the beekeeper, the exterior panels of the WBC provide additional insulation, allowing interior supers to be made with less thick wood, making them lighter and easier to lift. WBC hives hold ten frames.

TOP BAR HIVE

Also known as the *Kenya* hive, this model was created as a cheaper alternative to traditional hives. Owing to their low cost, these hives are more commonly used in developing nations. Top bar hives are physically quite different from other models. While they make use of the concept of bee space and use frames, the frames only have one side, the top, hence the name. When working with such frames, the beekeeper allows the bees themselves to create honeycomb without foundation support (sometimes foundation is used, but only as a tiny piece). As opposed to stacked wooden supers used in other hives, top bar hives extend horizontally. The structure is a trapezoid shape, instead of a cube, which encourages bees to draw the comb so that it hangs down instead of attaching it to the walls of the hive cavity.

While comb from top bar hives cannot be placed in a centrifuge for honey extraction (the comb would fall apart), the repeated buildup of new honeycomb after each extraction does create a surplus of beeswax that can be sold. Honey production from top bar hives is quite a bit less than that from other modern hive types. On the plus side, bees can be encouraged to store honey separately from brood, making it less likely that bees will be killed during comb removal, as is common with more rudimentary beehives like skeps. As cute as they might be, it is worth noting that it is illegal to keep bees in skeps in both the United States and Canada.

A top bar hive uses a trapezoid shape

WHERE TO PLACE YOUR HIVE

Once you've acquired your hive components and put them together, it's time to situate the fully assembled hive somewhere on your property. Several crucial variables should factor into your location decision, including sun exposure, windbreak availability, water source, flight path, weed suppression, accessibility, and moisture prevention.

Sun Exposure

The hive needs a bit of sunshine to keep it warm and active. A southeast exposure provides morning sunshine, stimulating foraging bees to rise, shine, and get busy gathering nectar. Siting your hive in this direction also warms the hive in cooler months while shielding it from the hottest midday rays of summer. Avoid placing the hive in direct sun, as that can warm the hive too much; a too-shady spot, however, may have problems with too much moisture.

Windbreaks

Bees need a bit of protection from heavy winds. In considering where to site your hive, be sure to keep it out of the direct path of wind gusts. After you've determined your spot to be relatively wind-free, place the hives with a windbreak of some form at the rear. This can come in the form of a hedgerow, a cluster of bushes, a fence, the edge of a forest, or, as in my case, the side of a building. If you're an urban rooftop beekeeper, you might want to consider locating your hives close to a wall, beside the roof access door, or near some potted shrubs to provide added protection from bracing winds.

Water

Access to water is crucial for a bee's survival, especially for temperature regulation inside the hive during warm months. Honeybees also use water to thin out any honey that may have become too thick. If you live adjacent to a natural water source such as a creek, stream, river, or pond, you're in great shape. Otherwise, consider a garden pond, birdbath, terra-cotta pot filled with water (and small rocks, so that the bees don't drown while drinking), or chicken waterer (what I use in the event the small stream located adjacent to my hives dries up, which happens periodically during summertime droughts). If you have nearby neighbors, establishing a designated water source right from the onset is crucial. You don't want your bees quenching their thirst in an unwelcome location, like your neighbor's dog bowl or koi pond.

Flight Path

Bees are creatures of habit. Accordingly, once they've set up a flight path out of the hive, they remain pretty steadfastly dedicated to it. When placing your hives, face them in the most unobtrusive direction possible. You want to face southeast, of course, but if your southeast happens to point directly toward a sidewalk, you'll need to provide some form of barrier. If your property line is close, plant a high hedge or put in a fence at least 6 feet (1.8 m) tall. The barrier will encourage the bees to fly up and away from anything in the immediate vicinity.

Weed Suppression

If you intend to site your hives in a grassy area, it will be important to stay on top of lawn maintenance. Keep the area in front of and on the sides of the hives free of weeds and grass. This prevents uninvited critters from catching a ride on a windblown blade of tall grass and jumping inside the hive for a look-see or a snack. Low ground cover also provides a free flight path for your bees.

Accessibility

Be sure to place the hive in a location that can be easily accessed. If it's crammed up against a building, moving around and behind it can be a challenge. Conversely, if it is placed in a wide, open field, far from any roads, getting there to do regular maintenance may be a chore. Site your hive where it can be both directly accessed with ease and where a vehicle, wagon, or other means of locomotion can reach it when the time comes for honey harvesting.

Moisture Prevention

While bees need water, they don't need it inside the hive. Excess moisture can be hazardous to bees. The hive's near-constant internal temperature of 90 to 95°F (32 to 35°C) causes water vapor in the air to condense, forming drops of water on the interior ceiling, which then drip down onto the bees, chilling and possibly killing them during colder weather. When choosing a site, avoid areas that are continually shady, that have perpetually wet ground, or that are near overhead water sources, like gutters or downspouts.

To remedy an overly moist hive, begin by placing the housing in an area that does not pool damp air. Tilt the hive forward slightly by placing 1-inch (2.5 cm) blocks of wood behind the two rear corners of the hive. This way, moisture will drop off and away from the hive. For additional winter ventilation, permanently glue four very thin, flat pieces of wood (such as craft sticks) onto the four corners of the flat side of the inner cover. A bit of additional airflow between the supers and the inner cover will help prevent condensation.

SETTING UP HOUSE

Once you've established where you'll locate your hive, determined that location meets all the criteria necessary for your bees to thrive, given your neighbors a heads-up, and examined local ordinances on beekeeping, it's time to purchase housing. A wide range of reputable beekeeping suppliers can be found worldwide (I've compiled a list in the Resources section on page 129). Order a few catalogs or peruse their websites, compare pricing, and consult with established beekeepers to decide which hive model and setup will work best for your purposes. My first two hives came as new colonies in separate deep hive bodies. I purchased an additional medium super for each so that the bees would have adequate storage space for food come winter. I make apiary purchases online or through a nearby beekeeper who keeps all manner of bee equipment on hand for purchase. He's about 45 minutes from me, so we rendezvous halfway when I discover I need something.

Housing is available for purchase either assembled or unassembled. If you're handy with carpentry and have time to spare before you need the equipment, unassembled hives and frames are a low-cost option. You'll pay about 10 percent more for preassembled housing. I was in a rush to get my bees into their new digs, so I opted for assembled housing when I was getting started. Depending on your level of woodworking abilities, it may be worth your while to do the same.

In most cases, your hive components will come without any paint or coating on them. If you've purchased woodenware, you'll want to paint all outer surfaces of the housing to keep it from rotting. Apply two coats of latex or oil-based paint to ensure that all porous surfaces are covered and no moisture can creep in. Choose light-colored paint, as anything too dark can cause the hive to get quite hot during warmer months. Paint only the outside of the woodenware. Frames never need to be painted or stained, nor do any surfaces that live inside the hive, such as queen excluders, wooden feeders, or the inner cover. It is worth repeating: the interior surfaces and woodenware should never be painted or stained.

Chapter 4
Feeding Bees:
Keeping the Hive Abuzz

Flower pollen and nectar are a honeybee's preferred foods. It's no coincidence that they are also the very best foods for them. As a bee happily gobbles up these plant secretions, it transfers pollen on its fuzzy legs, pollinating the plant. This symbiotic relationship permits bees and plants to simultaneously reproduce and flourish. On occasion, bees may fall short of meeting their food needs on their own. This is when you, their conscientious steward, will be required to step in, supplying food in order to stave off starvation of the entire colony. Bees can weather a great deal, but without sufficient food they'll perish, especially during colder weather. With a bit of aid from beekeepers, hives will stay well fed all year long, allowing them to not just survive, but thrive.

THE BIRDS & THE BEES

Many species of plants rely on other life forms to help them reproduce. A plant must be fertilized before it can produce seeds; this is achieved via pollination. Male components of flowers, called *anthers*, produce brightly hued yellow pollen globules. When these bits of pollen are transferred to the plant's female parts, called *stigma*, they unite with the egg inside the plant's ovary. Germination then begins. The flower grows, then dies back, drops its petals, and produces fruit and seeds.

While some plants are pollinated by wind or water, many plants rely on what is known as *biotic* pollination, getting by with a little help from their friends. In biotic pollination, helpful organisms known as *pollinators* transfer pollen from anthers to stigma; without assistance, the plant would not fertilize itself and could not reproduce.

Many different species of organisms may serve as pollinators and assist in this alchemical exchange, including vertebrates and invertebrates. Insects, however, shoulder the largest share of the work. The honeybee is one such pollinator among many; wasps, bumblebees, ants, beetles, birds, bats, hummingbirds, butterflies, moths, and more are vital in allowing pollination to occur. While all pollinators are critical, the anatomy of the honeybee is especially suited for the task. Tiny hairs on its body expertly gather up the sticky yellow orbs of pollen, transferring them far and wide as it buzzes about from plant to plant or moves up and down among the various parts of the same plant.

The helping hand (or rather, leg) that honeybees offer in pollination is invaluable. Up to one-third of all of the foods grown and consumed by humans are pollinated by honeybees. Bees' work in pollination is responsible for literally feeding our entire planet. Many people associate honeybees merely with honey, ignoring their indispensable work in pollinating so many of the crops consumed by humans and animals alike. Without the honeybee, the availability of many types of plant foods would be profoundly compromised.

A NEED FOR FEED

As they forage, busily pollinating along the way, honeybees gather up not just pollen but nectar, too. This transparent liquid excreted by plants is trucked back to the hive and used to make honey (we'll discuss honey production in greater detail on page 97). Honey provides the bees with carbohydrates, giving them the boost of energy they need to go about the multitude of tasks performed in their busy lives. The pollen brought back to the hive provides honeybees with fats, proteins, and vitamins. Pollen and nectar, along with water, compose the entirety of a wild honeybee's diet.

When plants are in bloom and the weather cooperates, honeybees are able to gather up all of the food they require without outside intervention. On occasion, however, a confluence of events can prevent them from acquiring nectar or pollen. Heavy rains, freak weather that kills off plants or compromises their ability to produce nectar, drought, robbing, an early, warm spring followed by a cold snap, and other situations can result in diminished honey stores for bees as they enter colder months. Without enough food to last through the winter, your colony runs the risk of starvation. In order to keep the hive alive, it may be necessary to supplementally feed your bees. The time of year determines what type of food you will provide them with. We'll examine food choices next, along with the types of feeding equipment available.

Knowing When to Feed

If you're new to beekeeping, you might wonder how exactly you'll know when supplemental feeding is required. By the end of the growing season the bees' efforts should have accumulated a honey stockpile of at least 45 to 70 lbs (20.4 to 31.8 kg). The variation in weight is based on your geographic location, with colder regions requiring larger stores of honey to overwinter your honeybees safely. If the larder is not sufficiently full, you will need to offer supplemental feed. Early spring and autumn are the two times of year when feeding may be needed.

There are two ways of assessing honey stores: *frame examination* and the *"heft" test*.

FRAME EXAMINATION

Prior to cold weather, examine each hive's frames. A shallow frame full of honey will weigh around 3 lbs (1.4 kg), a medium frame 4 lbs (1.8 kg), and a deep frame 6 to 7 lbs (2.7 to 3.2 kg). Add up the total weight to determine if your bees have enough food to last the winter.

HEFT TEST

Walk around the back of the hive and give it a lift from the bottom, checking its heft. You'll need to have a general idea of what 45 to 70 lbs (20.4 to 31.8 kg) feels like in order to perform this method of honey assessment properly. If you've ever hauled around a large bag of chicken feed, flour, or potting soil, you'll have a good idea of what such a weight feels like.

FEED OPTIONS

What you'll feed your honeybees is largely determined by the time of year. Warmer weather permits liquid feeding, while colder months call for solid food, as an excess of moisture may cause dysentery in winter. A responsible beekeeper is a prepared beekeeper, so keep supplies for making honeybee food on hand during early spring months and again as autumn approaches.

Honey

Arguably the best possible food for bees, many beekeepers keep several frames in their freezer all winter long, in the event that emergency early spring feeding is required. If you opt for this method of feeding, wait for a warm spring day to do it. Then, first scrape the *wax cappings* off of the honey-filled frame (wax cappings, more commonly known simply as beeswax, are what bees place over honey once they've filled a honeycomb cell, preserving the sweet, heady substance for future use). Bees form a tight ball (known as a cluster) during the cold months in order to trap heat. You'll want to place the uncapped frame of honey as close as possible to the cluster. If you have a strong colony, you can place the frame above the cluster in an upper super. If your colony is weak, you'll want to place the frame beside it, without actually touching or otherwise disturbing the cluster.

Only use honey frames from hives you know to be disease-free. While it may be tempting to purchase someone else's frames, it's not advised. Diseased honey could infect a hive already stressed by cold weather, resulting in avoidable deaths. Never feed your bees with store-bought honey. While purchased jars of honey are fine for you to eat, the stuff can be lethal for bees. Spores of dangerous AFB (American

Foulbrood, a potentially devastating disease, see page 85) are often found in such items, especially those produced by large-scale honey-production companies.

Sugar Syrup

Lacking a frame full of honey, sugar syrup is the next best bee-feeding option. This method only works, though, when the weather allows bees to take periodic flights outside the hive. If it is still too cold, bees won't break the cluster to reach the syrup. Workers will also need to leave the hive to relieve themselves, and liquid food requires more frequent flights for doing so.

Use only granulated white cane sugar to make sugar syrup for feeding. Don't use brown sugar, raw sugar, sucanat, molasses, or sorghum, as such products can cause dysentery in bees. For spring feeding, you'll want to use a 1:1 ratio of sugar to water. For autumn feeding, increase

the sugar to a 2:1 ratio. The thicker autumnal syrup creates a less moisture-laden feed, allowing the bees to render it into honey more quickly, preparing their cold-weather food supplies more expediently. To make the syrup, simply warm water before adding sugar, stir, and then allow to cool completely. If you're warming the water on the stovetop, do not allow the sugar mixture to boil. The sugar could caramelize, forming a somewhat indigestible and, at worst, toxic substance for the bees. Bear in mind that, when they really need it, they'll likely drink the sugar syrup dry in a day or two, so you may want to make a couple of gallons and refrigerate them up to three weeks until needed. The upcoming discussion on feeders presents a variety of options for making this sugar syrup readily accessible to tiny honeybee tongues.

Dry Sugar

During the chill of colder months, emergency measures can sometimes be necessary for a colony that has all but exhausted its honey stores. This can happen in late spring, when the queen has begun laying brood again and large amounts of honey are being fed to developing bees. Some beekeepers use dry granulated white sugar in such instances. Placed either directly around the opening of the inner cover or on a piece of paper laid atop the uppermost frames,

the sugar gives honeybees access to an immediate food supply. This feeding method doesn't always work, though. Sometimes the bees gobble it up straight away, other times they'll ignore it completely, and sometimes they carry it out of the hive completely, thinking it is waste in need of removal. If you opt to feed with granulated sugar, check back to see whether it is being consumed or not while doing so as gently, quickly, and discreetly as possible.

Fondant

Also referred to as *sugar candy*, fondant is another emergency food option. Simply a stiffened mixture of sugar, corn syrup, water, and often a thickening agent such as cream of tartar, fondant is formed into patties and placed directly on top of the uppermost frames. You can make the sugar candy yourself, order it from a beekeeping supplier, or even purchase it at a bakery (bakeries often use fondant in frosting cakes).

Put a patty of fondant inside a plastic freezer bag, roll it with a rolling pin to about a 1-inch (2.5 cm) thickness, and store it in the refrigerator until needed. When the time comes, score the bag with a knife on one side in an X. Then turn the panels of the X back, and place the bag on top of the uppermost frames. Place an empty super on top of the fondant, so that the bees can have adequate space to move freely around the patty.

Homemade Sugar Fondant

Making your own fondant is quick, easy, and inexpensive. All of the ingredients, if not already in your pantry, can be found at any grocery store.

YOU WILL NEED:
1½ cups water
2 cups granulated sugar
2 tablespoons corn syrup (organic, if possible)
⅛ teaspoon cream of tartar

TO PREPARE:

1. Combine the water, sugar, corn syrup, and cream of tartar in a small stainless-steel pot, and warm over medium-high heat. Stir gently with a metal spoon until the sugar is completely dissolved.

2. Once the sugar dissolves, discontinue stirring. Clip a candy thermometer to the side of the pot, and monitor the temperature until it reaches the medium-ball stage, around 238°F (114°C).

3. Remove the pot from the heat, and transfer the mixture to a shallow plate or dish. Allow it to cool until just warm to the touch.

4. Transfer the candy to a mold, such as a loaf pan, pie pan, or cake pan, and allow it cool until firm.

5. At this point, you can cut off a slice and take it to the hive, or store in a plastic bag in the refrigerator until needed.

Pollen

Densely rich in nutrients, pollen is fed to bee larvae and young bees. If a developing bee doesn't receive adequate pollen during the first few weeks of its life, deformities can occur. There is no perfect substitute for real plant pollen, so the foragers' gathering duties are absolutely essential for the next generation to thrive.

When flowers are in bloom, bees gather up large quantities of pollen, carrying it back to the hive in pollen baskets (*corbicula*) on their legs. In the hive, worker bees take the pollen from foragers and deposit it into honeycomb cells. As they fill the combs, the bees add a bit of saliva to keep the pollen from germinating, and cover each cell with a protective layer of honey and wax. This sealed pollen is referred to as *bee bread*.

During the early spring, when the queen begins laying brood once again but before flowers are producing pollen, the risk of running out of pollen stores becomes pronounced. If the colony has run through its supply, there are serious risks to developing and fledgling bees. During fall inspections (to be discussed in detail in chapter 9), you will want to check for pollen stores. Depending on the plant it was gathered from, pollen ranges in color from bright yellows and oranges to reds,

purples, and pale greens. Supplemental pollen can be provided by making pollen cakes or patties using pollen gathered from your own bees in summer. It is not advisable to feed your hives with pollen gathered from another hive, as there is the possibility of spreading disease.

If you don't have any pollen from your own colonies available, the next best thing is a pollen substitute. Available from beekeeping suppliers, pollen substitutes are a protein source containing all of the nutrients needed by developing bees without actually containing pollen. Once you begin supplying pollen, you will need to continue to do so until naturally available pollen appears on plants. If an interruption occurs in pollen access, it will adversely affect the rearing of brood and young bees. Similar to adding fondant, when adding pollen patties or pollen substitute to the hive you'll need to provide extra room for the bees to move around the food. Flip the inner cover over, so that the ledge on it faces down. Replace the inner cover, outer cover, and weigh down the whole thing with a heavy rock or brick to prevent the top from flying off during blustery weather.

Homemade Pollen Cakes

Provided you have pollen from your own colonies on hand, making pollen cakes is a relatively straightforward process. Soy flour can be obtained at natural food stores, as well as online through natural food purveyors.

<table>
<tr><td rowspan="4">YOU WILL NEED:</td><td>¹/₄ cup bee pollen*</td></tr>
<tr><td>1 ¹/₃ cups granulated sugar</td></tr>
<tr><td>³/₄ cup hot water</td></tr>
<tr><td>³/₄ cup soy flour</td></tr>
</table>

*If you don't have pollen from your bees on hand, substitute either ¹/₂ tablespoon brewer's yeast, 1 tablespoon powdered skim milk, or ¹/₄ cup soy flour. With brewer's yeast or powdered skim milk, you may need to add extra soy flour until you achieve a consistency resembling peanut butter.

TO PREPARE:

1. Combine the pollen and the sugar in a small mixing bowl. Add the hot water, and stir. Add the soy flour, stirring as the mixture thickens.

2. Transfer the mixture to a sheet of waxed paper. Place another sheet of waxed paper on top, and press the mixture to ¹/₂ to ³/₄ inch (1.3 to 1.9 cm) thick.

3. At the hive, remove one side of the waxed paper and place the exposed side of the cake directly on top of the uppermost bar, where the cluster is. Flip the inner cover over after placing the cake on, so that it accommodates the patty within the hive.

4. Check after one week, and replace with a new cake before the previous one has been completely consumed.

FEEDERS

A number of different feeder models for holding sugar syrup are available to the beekeeper. Differences are largely characterized by the material the feeder is made out of and whether the feeder is inside or outside the hive. Advantages and disadvantages exist with each model.

Boardman Entrance Feeder

This feeder model is one of the least costly options available. A wooden or plastic base holds a screw-top glass or plastic jug into which syrup is added. Some beekeepers, myself included, aren't fans of this model, for a number of reasons. The Boardman hangs from the side of the main bee housing, so in order to access the syrup, the bees have to climb off of the cluster, down the frames, and out of the hive. Another problem is that, owing to their location outside the hive, Boardman feeders can cause robbing if neighboring bees get a whiff of the syrup. Additionally, most of these models only hold a quart of syrup at a time. When you have bees in serious need of feeding, refilling the feeders can become a hassle. Furthermore, this model of feeder is vulnerable to theft or abuse by predators, such as raccoons or possums, who can easily swipe it off or pull it out.

While a similar feeder, made of plastic, contains a flat base that slides into the hive, allowing the bees to access syrup from inside, it too presents the same challenges of robbing and small capacity as the Boardman.

The Boardman does have a good purpose, though. Instead of being used as holders of sugar syrup, they are perfect for holding water. When the sun is bearing down during warmer months, the entrance feeder is an ideal vessel for an immediate water source. In extreme drought, a hive can easily go through a quart of water daily. Keeping water available in such close proximity to the hive allows you to closely and easily monitor their water use, replenishing as necessary.

Hive-top Feeder

Also known as *Miller feeders*, this model of bee feeder is made of either wood or plastic and covers the entire top of the hive. Variations exist, although the general design is the same: a shallow pan running the length of a super with screened areas in which honeybees can access sugar syrup. I use a hive-top feeder with my bees and find it enormously convenient. It rests atop the uppermost super and underneath the outer cover (no inner cover is needed with this model), so you are never in direct contact with the bees. This format enables the beekeeper to feed the colony without disturbing it, reducing the likelihood of being stung. Hive-top feeders also have a large holding capacity. Anywhere between 1 and 4 gallons (3.8 and 15.2 L) can be added at a time, depending on the model. Less frequent feedings are therefore required—great news for a beekeeper busy with other caretaking tasks in late spring and autumn.

The disadvantage of this model is that it is awkward to remove when full. Hive inspections can become potentially quite messy, especially if you opt for a plastic model, as it can wobble and spill syrup during removal. As such, I'd suggest planning your inspections in between feedings. Wait until the feeder is empty, check on your winged friends, replace the feeder, fill 'er up with sugar syrup, replace the outer cover, and rest contentedly.

Division Board

Also known as a *frame feeder*, this feeder model resembles a trough-like frame and rests alongside other frames directly within the hive. Made of either a single unit of molded plastic or wood, it replaces one or two frames. Accordingly, it allows for placement right beside the bees. Some beekeepers opt to keep division board frames inside their hives year-round.

Disadvantages to this model include the fact that the bees will be directly exposed to the weather when refilling. Cold-

Boardman entrance feeder

weather exposure can chill the bees, while hot-weather exposure can result in a robbing when neighboring colonies get a whiff of the sugar syrup. Another disadvantage is that fewer frames will be available for honey, pollen, and brood, as one or more must be removed in order to accommodate the feeder. Lastly, the design of division board feeders, with their open bin of syrup, can result in drowning for particularly overzealous bees. Some plastic models have interior walls with rough edges for bees to gain a foothold. Otherwise, if using a wooden model, you might add a U-shaped bit of metal hardware cloth over the top of the feeder to provide a base for your bees to grasp onto. A wooden dowel added to the feeder will achieve the same goal.

Feeder Pail

A feeder pail works by the formation of a vacuum, similar to that produced in a water cooler. Either a $1/2$- or 1-gallon (1.9 to 3.8 L) plastic pail is filled with syrup, covered with a lid perforated with six to 10 tiny holes, and placed lid-side down either directly over the uppermost frames (if the hive is weak) or over the opening in the inner cover (if the hive seems strong, but low on food supplies). Bees can access the syrup with their tongues, gathering up small beads. One of the biggest risks posed by the feeder pail model is that of leakage, an especially precarious situation if you're using this feeder in cold weather, as the dripping syrup could chill the bees. In order to assure that leakage doesn't occur, after placing on the pail's lid, turn the feeder upside down and allow it to drip either somewhere on the ground away from the hive (not directly next to it) or on the inner cover until it ceases to drip. Only then should you place it on the uppermost frames or over the inner cover opening. Place an empty deep super over the pail, then replace the outer cover, and top off with a heavy rock or brick to prevent the cover from flying off in windy weather.

Though inexpensive and relatively easy to use, the feeder pail does have its disadvantages. Because it will be placed either directly on top of the colony or over the inner cover, you will have to smoke the bees each time you refill the feeder. Smoking will put the bees on alert, increasing the likelihood for stings (see page 51-52 for a discussion on using a smoker). You know that your heart is in the right place and that you're just trying to help your pollinating buddies survive, but they don't know it. One way to reduce the risk for stings, if using the feeder over the inner cover opening, is to place a bit of closely woven hardware cloth on top of the inner cover, introducing a barrier between the feeder pail and the bees. They can still access the syrup, they just can't directly access you.

Plastic Bag Feeder

By far the most inexpensive feeder available, this model is exactly what its name indicates. All you need to do is pour some sugar syrup into a resealable 1-gallon (3.8 L) plastic freezer storage bag and then lay it over the uppermost frames. Using a razor blade, score the bag with several 1- to 2-inch (2.5 to 5 cm) cuts. Give the bag a gentle squeeze once you've made your cuts, allowing a bit of syrup to ooze out and entice the bees. Then place an empty super over the bag, topped by the inner and outer covers.

Plastic bag feeders, like hive-top feeders, reduce the likelihood of robbing, as the feeding is occurring inside the hive. The downside to this model of feeder is that, once cut, the plastic bags cannot be reused and must be discarded. Additionally, because the feeder goes directly on top of the frames, the beekeeper will need to interact with the bees directly, disturbing them and risking being stung. Opening the hive during cooler weather also runs the risk of chilling the bees.

Feeder pail

Opening a plastic bag feeder

Chapter 5
Essential Equipment:
Bee Prepared

Like any hobby, beekeeping has its share of accoutrements and supplies. In order to safely and successfully care for your winged charges, a bit of protective (and placating) gear is necessary. Some items, such as the smoker, are indispensable, while others serve more as handy, albeit nonessential, tools. Here we'll examine items both crucial and ancillary to the beekeeper. Each general category has multiple styles and models within it. Chat with fellow beekeepers about which makes, models, and styles they prefer, and provision yourself accordingly.

SMOKER & FUEL

The smoker is one of a beekeeper's most essential tools. Its use disorients the bees, permitting access to the hive. The technique may have been discovered when a prehistoric honey-seeker visited a wild hive by fire-lit torch one night and discovered the smoke seemed to induce a less focused (and ornery) hive. There are several reasons smoke has such a profound effect on bees.

As discussed in chapter 1, most of the communication between bees occurs via pheromones, scent-specific chemical messengers. If they perceive an attack on the hive, sentry bees emit an alarm pheromone, essentially sending out a silent, wind-borne SOS signal. The signal is picked up by other bees, who gather around their sisters in distress, scoping out the scene and stinging if the threat signal is perceived to be valid. Smoke blown into the hive interrupts pheromone communication, incapacitating the bees' ability to spread information. The bees then experience a short-lived breach in the chain of command, becoming confused. This interruption provides an opportune moment for the beekeeper to open the hive, examine it, and move out before order is restored.

In addition to disrupting pheromone communication, smoke also triggers some unusual behavior. Having evolved in the wild, dwelling in hollowed-out trees, honeybees recognize the scent of smoke as a potential threat of destruction by fire. Therefore, when smoked, a typical hive comes to a standstill. The bees gobble up honey, presumably in an effort to shore themselves with sustenance until a new home can be found. This gorging placates the bees, rendering them temporarily sluggish.

Centuries of encounters with honeybees can attest to the smoker's efficacy. While many incarnations have existed over the years, in its current guise, a smoker is a stainless-steel or copper canister, enclosed in a cage (to prevent the smoker from burning the beekeeper), with a raised spout and a bellows attached to one side. Kindling and fuel are placed into the canister and then lit. The bellows are blown vigorously, and the fire inside is allowed to burn until it smolders and produces a cool, white smoke. This is the smoke you are looking for. If you see any sparks, or your smoke pours out hot to the touch and in a thin, gray stream, then it's much too warm to apply to your hives. Wait for it to cool down, until it produces

a steady, thick, abundant white smoke that feels comfortable when puffed onto your wrist.

What should you use for fuel? Good choices include burlap, pine needles, cardboard rolled into a cylinder, bailing twine, dry leaves, and wood chips. My preferred manner of fueling my smoker is to roll a strip of cardboard up in burlap. (I source the burlap from a local coffee roaster, but it could also be found at a fabric store.) After adding a few broken up sticks of dry kindling to the bottom of the canister, I tuck in some balled-up newspaper, ignite it, and then, once I see flames, add in my cardboard roll. A few good squeezes of the bellows works to bring the flames up into the roll. I give my smoker about five to 10 minutes of continuous, cool smoking before suiting up and approaching my hives.

Give a puff to the hive entrance when you first walk up. Wait a few minutes, and then lift the lid and give another puff.

You don't want to suffocate your bees, merely to occlude their conversational abilities. Your smoker must stay lit for the entire duration of your examination. Be sure that it is adequately fueled and smoldering before you approach the hives. The smoker, when properly fueled, should be capable of producing smoke for up to 30 minutes. If you have multiple hives to examine, check between hives to be certain that you still have enough fuel. Better to refuel in advance than run out of smoke just when it is needed most! If the bellows on your smoker should ever give out or become damaged, it's possible to order a replacement part from beekeeping equipment suppliers.

SUITS & VEILS

A close contender for the "most valuable player" in the beekeeping equipment roster would be a beekeeping suit and veil. Wearing protective clothing while working with your hives helps assuage fears of getting stung. From full-body zippered jumpsuits resembling overgrown marshmallows to half-body jackets, beekeeping clothing comes in a wide range of styles and sizes. Bee suits are available with or without attached veils and are typically made of either cotton or lightweight polyester. The veils shroud and protect your head, permitting visibility but preventing access by curious (or cantankerous) honeybees. If attached to a suit or jacket, you merely zip the veil into place. Veils sold separately from suits are secured by wrapping drawstrings attached to the veil around the waist and then tying them in the front.

The rationale for protective clothing rests more on the fact that honeybees are notoriously curious than it does on any notion of aggressiveness. It is true that bees may act defensively if they feel the hive is threatened, which could occur if you jar the hive unnecessarily or overstay your welcome with too long an inspection. In most cases, however, any bees that land on you do so because they are curious. They are examining you just as you are examining them. Be mindful of exposed wrists or ankles, as bees especially enjoy dark areas and will climb into tight spaces to explore. Full-body beekeeping suits have elastic bands around ankles and wrists, while jackets enclose only the wrists. If you opt for a jacket (which is what I use), be sure to tuck your pant legs either into tall rubber boots or down into your shoes. Some beekeepers even affix tape around their pant legs and wrists to fully block bees from gaining access. Strips of Velcro or rubber bands will work just as well.

Tuck pant legs into boots or shoes to keep bees out.

Ready for action

The zipper prohibits a bee's entry into the veil.

No matter what form of protective clothing you choose, be sure it is light-colored. Generations of interactions with dark-furred forest creatures such as bears, raccoons, and other darkly dressed, hive-assaulting critters have created for honeybees an aversion to darkly colored moving things. Smooth-surfaced garments in white or other light colors placate honeybees, while dark or rough textures such as wool could trip the "danger" alarm if such surfaces are mistaken for fur. Play it safe and stay on the lighter end of the color spectrum.

GLOVES

Your protective gear will also need to extend to your hands. Long gloves made of leather or canvas are available for purchase from beekeeping suppliers. Basic rubber kitchen or latex gloves (such as those used in a medical setting) also work well. Many experienced beekeepers prefer to forgo the use of gloves completely, claiming they're cumbersome and prevent getting a firm grip on frames and other bits of equipment. I'd suggest that while a no-glove policy might be perfectly acceptable for someone with years of beekeeping experience, it's not the wisest choice for the novice. You don't need to be worrying unnecessarily about the possibility of stings on your hands while you're getting acquainted with feeding or trying to find your queen on a frame filled with thousands of bees. Take your time, use your gloves, and then, as you gain more confidence, decide if working with gloves or bare hands is the better fit for you.

HIVE TOOL

While it might look like it holds no special powers, the thin, metal hive tool is indispensable in beekeeping. As discussed in chapter 3, honeybees seal up any cracks and crevices in the hive with super-sticky propolis, blocking out bacteria, bugs, or any other hive interloper. Phenomenally viscous, propolis is no easy substance to pry apart. Enter the hive tool. Broad and flat on one end, the hive tool wedges expertly between all bits of woodenware, permitting you, the beekeeper, to gently force apart pieces that have become stuck together. A hive tool is also helpful for scraping away wax comb the

Hive tools help pry open hives and frames.

bees may have built in undesirable locations, such as between or on top of frames (this type of comb is referred to as *burr, wild,* or *brace* comb). The other end of the hive tool curves upward, enabling the beekeeper to pull up and otherwise manipulate frames. While you might be tempted to just use a putty knife or screwdriver instead, I can't recommend it. A putty knife lacks the strength of a hive tool, while a screwdriver's small head can damage woodenware.

Sturdy gloves are indispensible.

HELPFUL EXTRAS

Whereas the aforementioned bits of gear and equipment are essential to keeping bees both safely and properly, the following items, while not crucial, are quite helpful. Before rushing off to buy everything mentioned, chat with other beekeepers about some of their favorite tools. No two beekeepers are alike, and they'll undoubtedly have different opinions about what equipment is top tier. Solicit their feedback and then make an informed decision about what items you think would be best to have on hand.

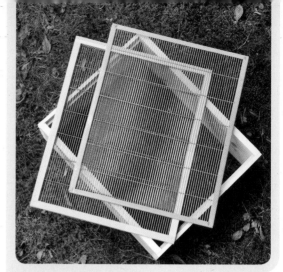

Queen excluder

Bee Brush

This long-handled, soft-bristled brush is useful for gently coaxing bees off any surface you'd prefer they not be on. Use the brush, or a long, soft feather, instead of your hand, which could accidentally squish bees. A bee brush is perfect for the delicate removal of honeybees from clothing and frames alike.

Queen Excluder

A wooden frame covered with perforated metal or plastic, the queen excluder is basically a "no queens allowed" restraint. The slats in the queen excluder are small enough for worker bees to move through with ease while prohibiting the larger queen from gaining entry. Placed between the hive body and the supers above it, the excluder confines the queen to the lower portion of the hive so that brood and pollen are compartmentalized separately from honey.

Come extraction time, all that will be in the frames will be the sweet, sticky substance, preserving the honey's clarity while simultaneously allowing all brood to develop.

Sometimes honeybees are reluctant to crawl up through the excluder and begin storing honey above until all available space within lower frames has been filled (they're quite the models of industry and thrift). To coax them upward, first place a super filled with empty frames intended for honey atop the hive body without the queen excluder. Once you've determined your bees are storing nectar in the upper chamber, insert the excluder. After you've extracted all of the honey you plan to extract for the season, remove the queen excluder.

Frame Holder

When a frame—shallow, medium, or deep—is filled with honey, brood, propolis, pollen, and wax, it gets heavy. Really heavy. Add to that the massive amount of buzzing,

active honeybees covering it, and you've got a weighty, delicate situation on your hands, literally. Assuage any fears about dropping your precious cargo by employing a frame holder. The frame holder does exactly what it says: it holds frames so that you don't have to. Two metal clamps grip onto the side of the super, with long shelflike arms extending out from the side. You simply remove a frame from the super and place it into the waiting arms of the frame holder. To see the other side of the frame, simply turn it around and replace it on the holder.

Frame Grip

Looking a good bit like metal jaws, a frame grip allows you to easily extract a frame from the hive with one hand. The grippers on each side of the spring-hinged handle grab hold of the frame firmly and securely. It is then quite easy to pull the frame from the super without harming any bees that might be moving over the frames. You can examine the frame with one hand, or move the frame to a frame rest for safekeeping.

Notebook

A small notebook can be enormously helpful for tracking the progress of your hives. Every time you visit, make notes about what you did, what you saw, and any impressions you have of the hive's health. Keeping a dated log of your actions can prove indispensable should problems arise.

Storage

Having all of your beekeeping supplies all together in one easily transportable collection streamlines the whole endeavor. A handled bucket or basket makes a good container. Stash everything in it, including your smoker, fuel, matches, veil, gloves, hive tool, bee brush, notebook, and more.

Take some time to read up on essential equipment options well in advance of acquiring your bees. You'll be that much better prepared to care for your new additions safely and properly.

Chapter 6
Obtaining Bees

After all of your research and planning, the time has finally come to get your bees! There are a number of methods of obtaining a hive of buzzing beauties. In the following pages we'll discuss package bees, nucs, established colonies, and swarms as possible means for honeybee acquisition. Each option possesses its own set of advantages and disadvantages. When I was a "new bee," I was completely lost when it came to this part of the beekeeping process. Thankfully, I had mentors and friends to advise me. Likewise, you have me right here with you, cheering from the sidelines, providing detailed information on each option. Learn what possibilities are out there and then make an informed decision about what seems best for you.

A TIME & A PLACE

As we've discussed in previous chapters, several key considerations need to be addressed prior to acquiring your first hive of honeybees. Before you welcome honeybees into your life you must first determine whether it is permissible to keep bees where you live, select the proper location for siting your hives, assemble all of the necessary housing and equipment, and assess whether you have the time and resources needed for keeping bees. Brush up on the subject of beekeeping by reading books on the topic (such as this one!), seeking out bee organizations in your area, and befriending or shadowing an experienced beekeeper. The beekeeping organization in my neck of the woods has been an absolutely invaluable resource. Through it, I've attended two years of bee school, connected with area suppliers of equipment and bees, participated in hands-on "field day" demonstrations, and met a woman who has become a cherished mentor.

After all of those preliminary measures have been attended to, the next step is to determine what type of bee you will keep. In chapter 2 we examined a number of the more popular species. Look over the list again, select what seems to be the best bee breed for your needs, and then begin looking for a supplier. The best time of year to get honeybees is late spring, before it becomes too hot to safely ship them. That said, you don't necessarily want to wait until then to begin your search, as most suppliers operate on a "first come, first served" ordering basis. During the quiet stretch of colder months (beginning even as early as November in North America), you can put in your order for honeybees. Then, once the warmer weather rolls around, your bees will be ready for pickup and installation into their new digs. Most suppliers begin shipping (that's right—bees can be sent in the mail!) or making bees available for pickup in mid-spring and continue on until June (in the United States, most bees are shipped from southeastern states; after June, the weather becomes too warm to permit safe shipping).

When the bees are due for arrival, be sure you've got their housing prepared to receive them. Much as you wouldn't wait until after anticipated houseguests arrive to tidy up the guest room and change out the sheets on the bed, don't wait until you've got a buzzing package of bees on your hands to order supers and a smoker. From the time they arrive, you've got about 48 hours to get them installed into the hive. Be a gracious host and greet your guests with proper hospitality.

"Going Postal"

If you opt to have your bees shipped to you (as opposed to using a local supplier and picking them up on location), it is imperative that you give your post office a heads-up in advance. They will call you as soon as your humming, buzzing shipment arrives. Be aware that post offices work through the night, and your package may very well show up at 3 A.M., so have the car keys handy! Tell the postal clerk to put them aside somewhere quiet and cool, hop in your car, and make a beeline for your new wards. You'll want to get your bees out of there pronto. Doing so not only keeps you on the good side of your postal staff, but also gets the bees out of the potentially inhospitable climatic conditions of the receiving area of the post office.

A PACKAGE DEAL (PACKAGE BEES)

Many fledgling beekeepers opt for package bees, for a number of good reasons. Package bees are exactly that: a package of bees and nothing else. Acquiring bees in this manner allows the novice to see the process of beekeeping from inception on through to care, maintenance, extraction, and beyond. If you are beginning with new woodenware, you'll witness its metamorphosis from empty foundation to drawn-out comb. Package bees are also small in number. Compared to the amount of bees present in a full hive (around 50,000 to 60,000), package bees permit the beekeeper to get a few notches in their apiary tool belt before advancing on to caring for the needs of an active, developed hive. If you're ordering package bees to place on frames with drawn-out, existing honeycomb, you can arrange to install the package in early April. Otherwise, if you will be adding bees to frames possessing only foundation, wait until it's a bit warmer to do so. Once the daytime temperature averages 56 to 58°F (13 to 14°C) (depending on where you are, late April to early May), you can then safely install your bees.

Available in 2-, 3-, and 5-pound (.9, 1.4, and 2.3 kg) packages (each pound comprises about 3,500 bees, so 3 pounds [1.4 kg] would fetch you just over 10,000 bees, perfect for starting out), package bees include a newly mated queen (cordoned off in a separate wire cage, often—but not always—with two or three attendant bees), a complement of worker bees, and sugar syrup for the bees to feed on in transit. Queens are isolated from the rest of the bees because, coming from different colonies, worker bees don't yet recognize her scent and may kill her before becoming accustomed to her pheromones. The whole package is about the size of a shoebox. Four sides are made from wood and two are covered with fine wire mesh screen for ventilation. The queen cage is approximately the size of a large book of matches. Her cage will come with a small amount of sugar candy for her and her attendants to consume while the other bees sip the sugar syrup.

You'll need to carefully examine your package as soon as you receive it. Look for a live queen and lots of moving, active bees. A bit of mortality in transit is normal, but any more than a half inch of dead bees on the bottom of the package should be reported. If your queen or large numbers of workers have died, either inform the clerk if you are collecting your bees from the post office or the individual from whom you are picking up your bees. Do so as soon as possible so that you may have your losses replaced. Package bees have undergone a good bit of stress during transit and should be housed, or *hived*, as it's referred to, as soon

as possible. Most suppliers won't deliver directly to you, so it will be necessary to keep them warm (but not hot!) while transporting them to their new home. If you intend to carry the package home in the open bed of a truck, cover it up with a loose-fitting cloth so that they don't get chilled, especially if the weather is cool. Otherwise, the interior of your car (minus the heat blasting at full throttle) is perfectly fine.

Once you get home, mist the package with cool sugar syrup. Not only does this give the bees a bit of food, it wets their wings, making it difficult for them to fly away (a good thing when you're attempting to move 10,000 bees into a new home). Next, move the package somewhere dark and cool (50 to 60°F [10 to 15°C]), such as a basement or garage, for about an hour. The bees will cluster around the queen and relax a bit. Have all your woodenware assembled and in its location of choice in advance. If at all possible, install the bees late in the afternoon or during the early evening hours. This limits the likelihood that any bees will fly away and lose their way home. It will be necessary to begin feeding your bees as soon as they've been added to the hive, and continue to do so until the first nectar flow occurs. A nectar flow, also referred to as a *honey flow*, describes the time period when nectar in flowering plants and trees is available for bees to consume. The flow periods vary widely, as do the nectar-bearing plants themselves, from region to region. To learn what is flowering in your area, check with your local beekeeping chapter or look online for information regarding seasonal nectar flows.

Two packages of honeybees

INSTALLING A PACKAGE

In order to successfully install a package of bees into an empty hive, you'll need the following items:

- A package of bees
- Hive body with frames
- Protective veil and suit or jacket
- Gloves (over time, you may come to feel gloves are unnecessary; in the beginning, though, I urge you to use them)
- Spray bottle of 1:1 sugar-water solution
- Hive tool
- A sewing needle, toothpick, or small nail
- Entrance feeder, hive-top feeder, or pail feeder

2 Wearing your protective gear, mist the package of bees with the sugar-water solution. You're trying to douse them, not saturate them, so don't overdo it when spraying. This step not only wets the bees' wings, making it difficult for them to fly away, but compels them to clean and groom the syrup off of each other, giving them something to occupy their time and energy.

1 Begin by removing 4 or 5 frames from the empty hive and setting them aside (you'll add them back in once you've emptied the package). Push the remaining frames all to one side. The empty space created will house the bees once you dump them from the package.

3 Holding the package in both hands, shake it down firmly against the ground. This loosens the bees from the interior of the package, causing them to fall to the bottom. Then, using your hive tool, gently pry open the flat wooden lid on the top of the package. Set it aside, but don't move it too far out of reach, as you'll use it again soon.

6 Shake any bees off of the queen cage and then examine it. Is the queen alive? If not, you'll need to call the supplier right away and request a replacement. You can continue with the installation of the other bees, though.

4 Inside you'll see a metal canister of sugar syrup with a strip of metal attached to its side. With the edge of your hive tool, pry up a corner of the canister. As you begin removing it, place a finger or two over the attached strip of metal; this strip is connected to the queen cage. Remove both the cage and the feeding canister at the same time.

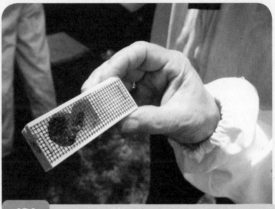

7 Queens are commonly shipped in small containers known as *Benton cages*. This style of queen cage consists of three circular sections cut out of a wooden block and covered on two sides with mesh screen. Two of the sections, or chambers, are for the queen to move around in, while the third contains a sugar paste (also known as a "candy plug"), made expressly for the queen to munch on in transit. You'll find corks on each of the shorter sides of the queen cage. Remove the cork from the candy end of the cage. Using a sewing needle, toothpick, or small nail, puncture the candy plug, taking care to not jab the queen in the process. The worker bees will use that puncture hole to begin chewing through to the queen over the next day or so. This gives them enough time to become accustomed to her scent and accept her. If introduced too quickly, before they've had time to acclimate to her, worker bees can sometimes attack and kill a queen.

5 Temporarily replace the wooden lid over the top of the box. Shake any bees from the feeding canister, and set it aside.

8 Carry the queen cage over to the empty hive body. With the candy end facing toward the sky, place the queen cage down in between the two center frames. Use the metal strip to form a hook over one of the frames, providing extra support. Some suppliers will use a short piece of plastic packing strap. In this case you will have to staple it to the top of a frame.

9 Give the package of bees another light spray of sugar water. Once again, firmly shake the package against the ground. Remove the wooden lid, and gently shake a cluster of bees over the queen cage. Next, shake the remaining bees into the area left by the removal of the frames. Place the package on the ground, beside the bottom board. Any remaining bees will vacate the package over the next 24 hours and fly into the hive.

10 Gently replace the frames, taking care to not squish any bees in the process. Given the opportunity and with gentle, slow, patient movements on your part, they'll shimmy out of the way when they feel and see the frames coming toward them.

11 Install your chosen feeding apparatus (hive-top feeder, feeder pail, or entrance feeder, see page 48) and replace the inner cover.

12 Put the outer cover on. Leave the bees to settle into their new space for about a week. When you return, check to be sure that the queen has been released from her cage. If so, remove the cage. Feed the hive continuously until an active nectar flow is occurring.

13 If the queen wasn't released from her cage, you'll need to release her directly into the hive. Remove the cage from the hive, take out the cork on the end opposite the candy plug, cover it with one of your fingers, and push the candy end of the cage—which by now should be free of candy, or close to it—into the front entrance to the hive. You're covering over one end so that she can only move out and into the hive, instead of moving toward you and flying up into the air.

A nuc of bees

(NUC)LEAR ENERGY

A nucleus of bees, more commonly referred to as a "nuc," is a miniature version of a hive. Nucs are comprised of a box containing 3 to 5 frames of bees in all stages of development. These frames will contain brood, baby bees, worker bees, food, and a laying queen. All you need to get started with a nuc is to remove the frames from their transporting box and insert them in a super.

The advantage of a nuc over a package is that the population of the colony will grow much more rapidly, as the queen's laying schedule will not be interrupted. A queen with brood, food, and a host of workers will have a wing up on the competition, so to speak. The entire hive will be that much further along the path of perpetuating itself. Bees in packages need time to draw out comb on foundation and will need up to three weeks before new bees are born. Nucs are also easier to install—simply put the frames into a super, and you're set!

As they contain more material (frames, brood, food) than packages, and are further along in colony development, nucs will cost more. That said, what is incurred in outgoing expense can usually be made up quickly. Given favorable weather conditions and consistent nectar flows, nuc colonies may have surplus honey available for extraction the first year you keep them.

One disadvantage to purchasing nucs is the potential for disease. Nucs do not have to undergo inspection or certification before being sold. As such, diseases or pests such as mites may be present. You can largely sidestep the risk of an unhealthy hive by purchasing nucs from a reputable supplier. Ask beekeepers in your community for recommendations. Local- and state-level organizations will be able to give you supplier evaluations, as well. If your nuc supplier happens to be nearby, contact a bee inspector and arrange to have the seller's hives inspected prior to purchase. You can find your local bee inspector by contacting your state or county extension agent or your local governmental agricultural agency.

SWARMING INTO ACTION

A third means of acquiring bees is through swarm collection. Swarming is a natural tendency of honeybees, serving as a means of propagation. When a hive swarms, it departs with the existing queen and about half of the colony, leaving new queen cells (these cells, which are slightly rounded and larger than those used to house worker bees, will be used to grow a replacement queen for the colony) and the remainder of the bees behind. Initially, the swarm flies around in a black cloud, eventually settling on a surface where it will temporarily rest while scout bees seek out a more suitable new dwelling. It is during this pause that you must act, if you wish to secure the swarm easily. Swarms can be an incredibly thrifty means of increasing, or establishing, your apiary. A swarm of bees contains somewhere in the

A swarm in action

neighborhood of 25,000 bees, including the queen. Compare that to the 10,000 or so bees in a package, and you've made quite the score! Of course, you'll want to make sure, to the best of your ability, that no one else has rights to the swarm. If any of your neighbors keep bees, do the neighborly thing and ask if any of their hives have recently swarmed before claiming the find.

Contrary to popular opinion, a swarm of bees is quite docile. Honeybees are protective of their homes and will act defensively to protect its contents. In the absence of an actual hive—and its attendant brood and food—bees are quite calm. Prior to swarming, bees consume about three days' worth of honey to fortify themselves until they can set up shop elsewhere, which satiates them. That said, while not difficult, capturing a swarm isn't really a job for a fledgling beekeeper. At the very least it would be helpful to have a seasoned beekeeper in tow to advise and assist you in the endeavor. An experienced beekeeper will also be able to determine if the swarm contains Africanized bees, in which case you should stay far away.

If a swarm alights in a convenient spot, such as a low tree branch or close-to-the-ground object like a picnic table, then capturing a swarm might not be too difficult. If the swarm lands in the upper reaches of a tree or other inaccessible location, however, then it might not be worth the trouble of removal. To capture a swarm from an area that can be reached safely, you'll need some kind of receptacle. Good choices include a cardboard box punctured for ventilation (poke about 20 tiny holes in each side with a barbeque skewer or ice pick), a super containing 4 or 5 frames of brood and food, or an empty nuc box. It's advised that less experienced keepers wear protective clothing for this undertaking.

If the swarm is hanging from a tree branch or bush, center the capturing box directly underneath it. Next, give the branch a quick, firm shake. This should dislodge the bees from the branch and down into the box below. Alternatively, carefully cut the branch off the tree or bush while holding it so that it doesn't fall to the ground once cut. Slowly carry the branch to the waiting box, delicately place it inside,

An experienced keeper capturing a swarm

close the top, seal it shut with tape, and then transport it quickly to its new home. If the swarm is hanging from a low-lying structure like a fence post or picnic table, place the capturing container underneath it, and then use a bee brush or your gloved hand to gently coax the bees downward. Inhibit their inclination to fly away by first wetting their wings with a light misting of water from a spray bottle.

You can hive the swarm into its new home several different ways. If you used a super with several frames of brood and food in it to capture your swarm, place the additional frames into the hive body once the swarm is in it, add a feeder and food (see page 48 for a review of feeders), and replace the outer cover. If you used a nuc or cardboard box for capture, you'll want to have a hive body with frames and foundation or drawn comb set up in advance for receiving them once you return home. You'll be feeding the bees as soon as you hive them (unless there's an active nectar flow happening), so have your feeder of choice in place, filled with sugar syrup, as well. Then, using either a bed sheet or a long sheet of plywood, create a ramp leading from the ground up into the hive entrance. Next, shake the honeybees directly onto the ramp as close to the entrance as possible. Compelled by the scent of the comb and food, coupled with pheromones emanating from their fellow hivemates and queen, the bees will begin a steady march upward into their new home. After two to three days, take a peek inside and look for the queen. If you can't find her, look for indications that she's there, such as eggs and capped brood. Absent those signs, you might have a queen-less hive, in which case you'll need to purchase and add a new queen as quickly as possible.

Profile of a Beekeeper

Chase

Tucked far from his native Manhattan, Chase tends his 10 beehives on 20 acres in rural New England. The Regional Director and Director of Sustainability for the Princeton Review, he cares for his mix of Italian and Russian bees from the home he has been updating and repairing with "super energy-efficient" and "green" technologies. Once he acquired the land, small-scale farming and beekeeping seemed like logical steps. Now in his fifth year of stewarding bees, he practices what he refers to as "plug-and-play" beekeeping. He describes himself as a "hands-off beekeeper, not using any chemicals and not messing with them much at all, other than feeding when necessary. The way I see it is that if the queen is strong, the hive will thrive. If not, well…that's life on the farm, I guess." Chase opines that "Nature tends to always know better than we do, so let the bees do their thing. Too many beekeepers overanalyze and obsess. Every time you crack open the hive, you stress them. The only time to intervene is when there are obvious signs of disease."

Where he has been proactive, as far as his winged friends are concerned, however, is in erecting bear fencing. After two years without any bear activity, year three proved quite different. A determined bear came through when honey stores were high and decimated three of Chase's hives. He immediately countered the bear's actions by purchasing a chain-link dog kennel kit, which he erected around his remaining hives along with a solar electric fence perimeter, nail boards, and razor wire. While the set-up was "a bit of an investment," Chase feels it was money well spent, claiming it will "last basically forever, and if some bear actually got through all that, they deserve the prize." Aside from this apiary version of Fort Knox, his hands-off approach seems to have benefited his hives nicely. His seasoned advice: "If set up properly at first, and fed as needed, many hives can just do their thing without intervention. Leave 'em bee, so to speak."

THE ESTABLISHED ORDER

A final option for obtaining bees is to purchase fully established colonies. This setup involves acquiring full hives of bees (supers, frames, bees) from a local beekeeper, often one who is moving, retiring, or needing to unload an abundance of bees from busy hives. While this arrangement can give you a leg up on establishing a beekeeping operation, it can be a bit daunting for the beginner. You suddenly have a thriving, buzzing, busy mass of 50,000 to 60,000 bees on your hands, possibly needing food, additional supers, and a bit of skill where their care is concerned.

As such, jumping into beekeeping in this manner is usually not recommended for first-time beekeepers. It can be daunting to encounter so many bees all at once. Additionally, some of the subtleties witnessed when starting from scratch with new woodenware will be lost, including viewing comb formation, the creation and later capping of brood and honey, and introducing a queen. Full hives of bees could potentially harbor diseases and pests, and used woodenware may be antiquated or worn-out.

That said, when I got started with bees, I elected to dive right into the deep end, purchasing not just one, but *two* full hives. I'd devoured a mountain of books on beekeeping, attended two weekends' worth of bee school, and secured a host of mentors. A tip from a fellow beekeeper about a 30-year beekeeping veteran wanting to sell off some of his many hives for a bargain sealed the deal for me. My bee buddy Jenny and I grabbed our gear, her truck, and an intrepid spirit and headed two counties over to secure my hives. We waited until most of the bees had come home for the night, sealed them up, and then moved them, slowly, to the back of Jenny's truck. Arriving at my house past dark, we carefully, steadily moved them to their new location in my fledgling bee yard.

Admittedly, it *was* daunting opening the hives the first time. Fortunately, Jenny came back around to assist with adding supers, and another local beekeeper came by for a full hive inspection. I'm a quick learner, and I always love a challenge, so my decision to start with established hives worked well for me. Only you can decide what feels comfortable to you. Assess all of the ways of obtaining bees carefully, and make the decision that best suits your needs and concerns.

SUPPLY & DEMAND

So, you've considered all of your procurement options, selected the breed of bee you want, checked in with the neighbors and the local authorities, gotten all of your woodenware and gear together, and are now primed and ready to take on bee stewardship. Where to find all of those packages and nucs and established colonies? Depending on where you live, it may be possible to source bees from a nearby supplier. Check with your local beekeeping organizations for referrals. Your local newspaper may have classified ads offering bees for sale, as well. If you're dealing with an individual, as opposed to an established company, have the government-appointed bee inspector for your area check the prospective hives for diseases and pests before purchasing. Otherwise, seek out reputable suppliers who will ship bees or make pick-up arrangements with you. You can find many by simply conducting an online search. There are also a number of fantastic periodicals advertising trustworthy companies (see Resources on page 129 for periodical and supplier listings). As with any business, some honeybee suppliers will offer better customer service and "product" care than others. Ask your fellow beekeepers and local organizations for suggestions about who to trust.

Profile of a Beekeeper

Majora

Working from her home and office in the South Bronx, Majora is an urban beekeeper and green economic consultant on a mission. Her enterprise, Majora Carter Group, by its own description offers "pioneering solutions to concentrated environmental problems that are grounded in a progressive economic development approach." One successful project aids those who face barriers to employment due to a lack of critical employment skills or a history of incarceration. Now the program trains 60 students per year in "green collar" jobs such as urban forestry management, green roof installation and maintenance, brownfield remediation, and wetland restoration. The way in which her business model relates to the honeybee involves the creation of a job training and value-added supply chain for recently released convicted New Yorkers, similar to the Sweet Beginnings endeavor based out of Chicago.

From her urban setting, Majora currently keeps three hives. She'll be making splits soon, increasing her apiary size to six. Her hives are currently comprised of Italians, but she'll be adding Russian queens to the mix to help in diversifying the gene pool. Majora's honeybees benefit the greater Bronx area, helping pollination efforts and the lives of those looking to move toward a better career path alike. As for advice on caring for the bees themselves, Majora advises a live-and-let-live approach. "They pretty much take care of themselves, so don't mess with them unless you have to."

Chapter 7
A Look Inside

You've acquired your bees and situated them safely inside the hive. Now it's time to open it up and take a peek. Much like any house visit, a bit of etiquette will go far in engendering goodwill toward your hosts. You want to get invited back, don't you? Here we'll cover the proper way to visit your hives, as well as what to be on the lookout for during inspections.

BEING A GOOD HOUSEGUEST

It's important to keep in mind that bees have schedules just like we do. They also have preferences when it comes to being called upon. As such, there are a number of dos and don'ts when it comes to visiting your hives. Knowing when it's time to say "howdy do" and when to say "perhaps another time" are essentials in being a good houseguest to your bees.

Do's

- ◻ Choose a warm, sunny day.
- ◻ Visit during the hours of 10 A.M. to 6 P.M., when a large proportion of the colony will be out foraging.
- ◻ Check in once every two weeks in early spring and mid to late autumn.
- ◻ Move gently and quietly; fast, brusque movements put the hive on alert.
- ◻ Practice good personal hygiene around the bees; heavy sweat or body odors are not appreciated in the bee yard.
- ◻ Wear your veil. Even if it's the only protective garment you decide to use, always use it.
- ◻ Wear gloves or, if going gloveless, take off your rings, in the event that your hand gets stung and you're unable to remove them.

Don'ts

- ◻ Come by on rainy, windy, or cold days.
- ◻ Overdo it with the smoker; a gentle puff is enough to give the hive a sense of what's going on.
- ◻ Inspect excessively or unnecessarily.
- ◻ Linger; the bees will be fine with a short visit, but overstay your welcome and they'll be less than pleased.
- ◻ Drench yourself with perfume or cologne before visiting the hive; the scents can attract bees, which you don't want (you're trying to remain as inconspicuous as possible, remember).
- ◻ Breathe heavily on the hive; it's apt to make them ornery. Whatever you do, don't eat a banana and then breathe on the hive; bananas smell like the bees' own alarm pheromone, and you might inadvertently sound the "danger" call.

INSPECTOR GENERAL

In order to perform seasonal inspections and attend to your hives properly, you'll need to know what exactly it is that you're looking for each time you open the hive. Bees neither like nor benefit from unnecessary visits, so being prepped on what to look for benefits you both. Here are step-by-step instructions for how to open the hive, how to remove a frame, what to look for on those frames, and how to replace the frames and put the hive back together.

Profile of a Beekeeper

Jon

Jon might have first picked up the beekeeping buzz years ago, from a father who kept a few hives (jokingly described as more of a "bee haver" than a "bee keeper"). That initial introduction would mature into a career centered chiefly around "everything bees." After forays in construction, long-haul truck driving, and commercial fishing in Alaska, he set his sights on all things *Apis mellifera*. From beekeeping (he currently tends to approximately 65 hives) to bee supplying, bee removal, bee education, and more, Jon's range of expertise and honeybee stewardship is vast.

Currently in his sixth season as a beekeeper, Jon is a trusted and highly regarded source of bee information, knowledge, and advice, not to mention beekeeping equipment in his community and beyond. A love of the outdoors, nature, and ecology coupled with a desire for self-sufficiency on his steep, rocky land made keeping bees the ideal career choice for Jon. Plus, as he puts it, "It just plain feels right." With so many hives under his care, he's weathered his fair share of successes and failures. According to him, that's all part of the requisite beekeeping learning curve. "No matter how good you are, you will eventually be humbled, be it by bears, mites, weather, or forces unknown. Perseverance and the pursuit of more knowledge and experience is the only key to becoming a successful beekeeper."

OPENING THE HIVE

1 Put on your protective gear. Ignite your smoker (see pages 51-52 for instructions), and be sure it is well-lit and producing cool smoke.

2 Approach the hive from the side or the rear, never from the front (doing so will interfere with the bees' flight path, which could irritate them). Stand to the side of the hive's entrance and gently direct two or three long, full puffs of smoke at it. Wait three to four minutes for the smoke to take effect.

3 Move to the rear of the hive. Lift one corner of the outer cover, and direct two or three puffs of smoke inside. Wait a few seconds for the smoke to take effect.

4 Set your smoker on the ground or use the curved end of it to hook it to the super's edge. Remove the hive's outer cover, exposing the inner cover or hive-top feeder, if you are using one. Turn the outer cover upside down, and place it either on the ground or on a raised object directly adjacent to the hive. You will be stacking other woodenware on top of the outer cover.

5 If you are using a hive-top feeder, give a puff of smoke to its outer edges, where the bees gather to feed at the screened areas (only remove a hive-top feeder when it is empty, or you could end up with a big mess on your hands). Otherwise, you'll be removing the inner cover. Using your hive tool, gently wedge it under one corner of the inner cover and the super below.

6 Give another puff of smoke under the inner cover. Moving slowly and carefully, pry off the remaining corners of the inner cover. Lift up the inner cover, and look to see if the queen is there. If you see her, gently brush her back down into the supers below. If no queen is present, using two hands, remove the inner cover, bees and all. Place it diagonally toward the entrance, so that the bees still clinging to it can walk back inside.

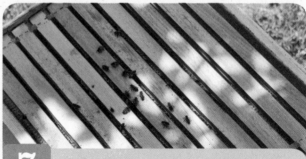

7 The hive is now open for business. Give a few, long, full puffs of smoke into the super, across the bees. You'll see them start to move down into the lower supers. Time to start removing frames and determining what's going on inside the hive!

REMOVING FRAMES

1 You'll be inspecting frames from one end to the other side. Always work in a clear, chronological order. The bees arrange things very specifically on the frames, and you need to be mindful of that.

2 Using the curved end of your hive tool, gently wedge it between the first and second frames. If any bees are in the way, a gentle nudge with the hive tool is all that's needed to get them moving.

3 Move the hive tool slowly from side to side, loosening the first and second frames from each other. Do the same thing on the opposite end of the same two frames.

4 The frame should now be loosened. If you find any burr comb between the frames, use the flat end of your hive tool to scrape it off. I bring a glass wide-mouth canning jar with me on my inspections and scrape any burr comb I come across into it for later use in beeswax-making projects. Using both hands and making certain you have a firm grip and that no bees are under your fingers, lift the frame up and out of the super.

8 Continue moving through all of the frames in this manner, working thoroughly but quickly. As you move closer into the cluster of bees in the center, be careful that you don't allow your queen to fall to the ground. Hold each frame over the hive opening when you are inspecting those frames, so that any bee falling off will fall right back down inside.

5 Gently rest the frame vertically on the ground beside the hive, inside of the overturned outer cover. It will most likely have bees on it, and that's perfectly fine. Again, mind the bees, taking care not to crush any as you put the frame down. Alternatively, if you've got a frame rest, you can mount the first frame on there.

9 You'll most likely need to give a few more puffs from your smoker midway through your inspection, which is why you want to be certain your smoker is going strong right from the onset.

6 You've now created an open area, which will enable you to access and inspect the remaining frames with greater ease. Remove the second frame in the same manner as the first, only this time don't put it on the ground or right into the frame rest.

10 Ideally, your entire inspection should take no more than 10 to 15 minutes total. If it takes you a bit more than that in the beginning, don't worry. You'll get the hang of it over time, and the hive will let you know when they've had enough of you for the day by an increased pitch in buzzing.

7 With the sun behind you, examine the frame on both sides, turning it vertically (end-over-end) as you do so, remembering to hold it firmly. Again, if you're using a frame rest, you can simply put the frame in the frame rest, examine it on one side, remove it, turn it vertically end-over-end, and place in the frame rest to examine the opposite side. Once you've examined the second frame, either leave it in the frame rest, if using, or place it beside the first frame, similarly resting vertically on its side on the ground. If you have chosen to use a Top Bar Hive (see page 39) that does not use frames, you will need to keep the comb hanging vertically during your inspection or it may snap right off the top bar.

WHAT TO LOOK FOR

So, then, what exactly is it that you'll be looking for? Essentially, everything you're concerned about relates to the queen. What she is or is not doing will be evident via a trail of visible indicators. Those are what you're after.

1 Begin by looking on a frame for the queen herself. She will most likely not be on any outer frames, but you never know! If you don't find her, start looking for her "signs."

2 With the sun to your back, look on a frame for the presence of eggs. These resemble tiny grains of rice and should be placed one per cell. Their presence means the queen is alive, or at the very least that she was there as recently as two days ago. Ideally you want to find eggs every time you inspect. While egg-laying will taper off and ramp up at different times of the year, eggs should generally always be visible when you're inspecting (laying ceases during colder months, but then, you shouldn't be inside the hive inspecting it to find that out!).

3 As the eggs grow, they move through various stages of physical development (see page 21). During the larval stage, you'll see white grubs in the cells. Once they mature into the pupae stage, the grubs will be capped over by nurse bees. Referred to as "capped brood," this stage is evidenced by the appearance of dark, tan-colored wax.

4 The pattern that the capped brood appears in is quite telling. Ideally, it should be in a "rainbow pattern," meaning that the middle of the frame should hold capped and uncapped brood, with each cell filled, followed above it by foodstuff, including pollen, royal jelly, and both capped and uncapped honey. The colors should all progress in a gradient fashion, much like a rainbow. If the brood pattern is spotty, spread out, and missing in lots of holes, you might have a problem on your hands. Furthermore, the cappings on the brood should be smooth, glossy, and curling slightly outward. If they are curling inward or otherwise look ragged, this could be another indication that your hive's queen may be ill or simply failing from old age. We'll discuss several means of handling this situation, should it present itself, ahead in Requeening (see page 79).

5 During spring inspections, and to a lesser extent in summer, you'll also be looking on frames for indications of the presence of queen cells. These will show up either as swarm or supersedure cells, or both. Both cells look an awful lot like peanuts—bulbous and slightly elongated. *Swarm cells*, usually visible on the bottom bar of frames (although sometimes simply in the lower half of the frame itself), indicate that the hive has become too crowded or that it is too stuffy inside in order for it to function as it should. *Supersedure cells* indicate that the colony has decided, for one reason or another (including old age, disease, or injury), that their queen needs replacing. These cells will most often be situated on the upper portion of a frame. The presence of either type of queen cell usually calls for a response on your part. We'll examine dealing with both situations on page 79, under spring management practices for your hives.

REPLACING FRAMES

3 Return the first frame last. If any bees remain on it, give one edge of it a sharp rap onto the inner cover resting in front of the entrance.

4 Make sure all the frames are placed both equidistant from each other (the rabbets on their ends will do this bit, provided you've pushed them together snugly) and from the outer walls of the super on both ends. This is very important. If you leave unequal distances, the bees will gladly fill it up with burr comb, making expedient access to the interior of the hive less likely during your next inspection.

5 Once the frames are replaced, it's time to close up the hive completely.

1 Push the frames back together in the order you first encountered them. Make sure they are also facing the same direction as they were when you first removed them. If you turned them over end-to-end vertically, move then back through the same movement sequence until they face their original positioning.

2 Push the frames together as a single unit, instead of pushing each one back into place individually. You're much less likely to accidentally squish and kill any bees this way.

CLOSING THE HIVE

2 Grab hold of the outer cover and gently slide it into position over the inner cover, making sure you don't cover up the ventilation notch. This is easily achieved by pushing the outer cover, once atop the inner cover, all the way forward.

3 That's a wrap. Everyone and everything is as it should be!

1 Once you've returned all of the frames, return the hive-top feeder, if you're using one, and refill it. Firmly knock any last bees off of the inner cover by giving it a firm rap on the ground. Slide the inner cover on, taking care to move slowly from rear to front so that any bees can move out of the cover's way. If your inner cover has a ventilation notch cut out, it should be facing upward.

Chapter 8
A Year of Bees

The needs of your bees change with the seasons. Some periods will be full of activity, both for the hive and for you, the beekeeper. Others will be characterized by dormancy and rest. This chapter offers detailed checklists for what care the hive needs during different times of the year. Being prepared for meeting the bees' needs will go far toward ensuring the success of your hive.

SPRING

Early spring is characterized by fluctuations between warm and cold weather. In the same week, there could be a series of sunny days and blooming flowers followed by a burst of snow. This extreme weather variation occurs when the hive is at its most vulnerable. The warm periods and longer days prompt the queen to begin laying eggs again. If brood production occurs quickly, the colony may eat through its food stores before new nectar sources are available or consistent in supply. Your job during the spring months, as the hive's steward, is to watch carefully for starvation, swarming, or external threats such as robbing or predation. Under your supervision, if your hive was strong going into the winter and was provided with adequate food, your winged friends should make it through spring without a hitch. On the following pages you'll see some important spring-specific tasks you'll want to consider addressing with your bees. Find complete seasonal checklists for the entire year on page 128.

Checklist

Early/Mid Spring

☐ Look inside only on warm, sunny days.

☐ Check for eggs, uncapped and capped brood, and general indication of a queen.

☐ Begin supplemental feeding if food stores are low; continue until an active nectar flow occurs (remember, nectar flows will vary from one geographic location to the next; if in doubt, try to connect with other beekeepers in your area to learn when nectar flows occur).

☐ Remove mouse guard and entrance reducers.

☐ Install queen excluder, if using, over brood box.

☐ Add supers as soon as brood box is full of bees.

☐ Continue adding supers as each new one fills with honey or bees.

☐ Begin weekly inspections in mid-spring, checking for swarm cells. See page 78 for measures to take for swarm prevention.

☐ Remove old frames and replace with new frames and foundation if equipment looks worn.

☐ Make splits as needed (see page 76).

☐ Requeen or allow natural supersedure as necessary (see page 79).

Mid/Late Spring

☐ Reverse hive bodies (see page 76), as this may act as a deterrent against swarming.

☐ Determine what swarm-capturing method you will use, in the event that one should occur.

☐ If an intense nectar flow happens, a late-spring honey extraction may be necessary.

☐ Plant honeybee-loving plants near your hive and around your property (see page 95 for some suggestions).

☐ Examine varroa mite population and treat as desired (see page 87 for a discussion of varroa mites).

Reversing Hive Bodies

Reversing hive bodies is exactly what its name indicates: switching the positions of the hive bodies, moving the one on top to the bottom. This is done to prevent swarming, as bees congested up near the top of an upper hive body (where they snuggled together to cluster for the winter) are unlikely to move back down below. Wait until a clear, warm day to do this. Reversing hive bodies when the weather is too cold could result in brood becoming chilled in the lower hive body as the queen and worker bees move upward into the upper hive body to being laying more eggs. Of course, if you wait too long, the hive could swarm. The timing of reversing hive bodies, is, admittedly, a tricky dance, but you should have no problem if you follow these steps.

1. Don your protective gear and light your smoker.

2. Lift the outer cover and give the hive a gentle smoking. Place the outer cover on the ground.

3. Using your hive tool, pry open all four corners of the upper hive body. With the inner cover in place on top of it, gently place the hive body on top of the outer cover.

4. Remove the lower hive body. It should be mostly empty of bees. Place it across the first hive body you removed, facing the opposite direction.

5. Take this opportunity to clean off your bottom board. Use the flat end of your hive tool and scrape out any wax or dead bees that may have accumulated. Be sure to hang on to the debris, though, as leaving it on the ground may entice predators to come snooping.

6. Now, take the hive body that is mostly empty (the one resting on top of the two hive bodies on the ground) and place it temporarily on the ground. If you have an empty hive stand or a level object like a cement block available, place the hive body on top of it.

7. Place the second hive body, the one that was formerly on top and is full of bees, in position over the bottom board.

8. Give the new lower hive body (the one full of bees) a gentle smoking, and use your hive tool to pry off the inner cover.

9. Take the hive body you have placed to the side (the mostly empty one) and put it on top.

10. Replace the inner and outer covers. All finished!

NOTE: If you are using a configuration of woodenware other than double-stacked hive bodies (I'm thinking here of a friend who uses all medium supers), reversing hive bodies is performed in the same manner as just described. Simply switch out the uppermost super with the lower one, following all of the steps as detailed.

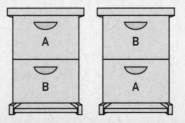

Making Splits and Nucs

Although essentially the same thing, in bee parlance, *splits* or *making a split* are terms used to describe dividing a hive for one's personal use, whereas *nucs* refer to divisions of bees one intends to sell. Making splits or nucs are both means of reducing the size of a hive that has become too congested and may be exhibiting signs of an intention to swarm. Divisions should only be performed on strong, thriving hives.

In your first year of beekeeping, it is highly unlikely that the size of your hive will increase to the extent that making splits or nucs is necessary. However, in subsequent years, such measures may need to be taken. If you are nervous about making splits or nucs on your own, I'd encourage you to either shadow an experienced beekeeper as they make divisions in their own hives, or ask them to accompany you in your bee yard. The best time to make a division of any sort is in early spring, once it warms a bit but around a month or so before the first nectar flow. Here's how to do it:

1. Once you've determined that you have a strong enough colony to merit division, put in an order with your preferred beekeeping equipment supplier for all of the makings of a new hive: bottom board, hive body, supers, frames with foundation, feeder, and so on. If you're making a nuc, you'll need the woodenware for a nuc box.

2. Next, place an order for a queen. Alternatively, if swarm or supersedure cells are present, you can allow the hive to produce its own queen (see page 79).

3. Once your queen arrives or you find capped-over queen cells, it's time to make a division. Position your new woodenware where the new hive will be situated. It needs to be at least 3 feet (.9 m) from the original hive.

4. Light your smoker, lift the outer cover, and give the hive a gentle smoking. Place the outer cover on the ground.

5. Look through the occupied hive body and find a frame with some honey and pollen on it. Place this frame in the empty hive body, or empty nuc box if you're making a nuc.

6. Locate the frame containing the queen. Move this frame into the empty hive body or nuc box. If you see any other queen cells on this frame, remove them.

7. Look for two additional frames containing pollen, honey, and brood in all stages of development (eggs, plus capped and uncapped brood). Place these frames into the empty hive body or nuc box, along with all of the bees on them.

8. Fill up the remaining spaces in the new hive body with empty frames. Fill your feeder of choice with sugar syrup, put it in place, and put on the outer cover.

9. Return to the original hive. It will contain queen cells but no queen. You can now do one of two things: introduce the new queen you've ordered, or allow for a virgin queen to emerge from a supersedure cell.

For natural supersedure, remove any capped or just-about-to-be-capped queen cells (these are bulbous, rounded cells with openings at their bottoms or sides, typically located at the upper portion of a frame) from the original hive, leaving several that have visible larvae in them. Push the remaining frames together, fill the empty spaces with new frames, install your feeder, and put on the outer cover.

If you will be introducing a new queen, push the remaining frames almost together, leaving enough room in between the two innermost frames to hold onto a queen cage. Review the steps outlined on page 59 for installing a package, and position the queen cage between the frames with the candy end facing upward. Fill the remaining spaces with empty frames, install your feeder, and put on the outer cover.

10. Whichever method of generating a new queen you choose, return after one week. If using a purchased queen, check to be sure that she has been freed from her cage. Remove the empty queen cage and continue feeding until the first nectar flow. If allowing for natural supersedure, remove all but one of the now-capped queen cells (it's smart to leave the first capped cell you find and move on looking for others; it might just turn out to be the only one in the hive, and you'd hate to have destroyed what would be your only available queen!). Close up the hive and leave it for one month, continuing to provide sugar syrup as needed.

11. You've now successfully created two hives from one! Either monitor the new hive, providing sugar syrup and supers as needed, or sell your nuc box to another beekeeper.

Preventing Swarming

Sometimes, no matter how hard you work to prevent it, your hive will swarm. It's the bees' genetic tendency—they're literally doing what nature tells them to do. That said, the following measures will go far toward thwarting swarming. Again, they're not foolproof, but if swarming is something that you absolutely want to prevent (some beekeepers feel more strongly about preventing swarming than others), then taking these steps adds an extra bit of insurance to your goal of keeping the bees right where they are.

→ Add supers as needed. Be attentive to how much space the bees have. If they feel crowded, they're more inclined to secure new digs. Beat them to the punch and give them the extra space they need before it's too late.

→ Add a queen excluder. Do this in early spring, before nectar flows might prompt swarming.

→ Reverse hive bodies. Bees like to move upward. If they're gathered in a cluster in the upper portion of the hive body and the queen resumes egg-laying, they might soon run out of room, even if a lower hive body has plenty of vacant space. By reversing hive bodies, moving what's on top beneath and vice versa, you provide "upstairs" rooms for storing food, brood, and more. Do this in early spring, before the first nectar flow. See page 76 for details on how to reverse hive bodies.

→ Provide proper ventilation. If the hive feels too hot to the bees, they'll feel inclined to search for somewhere a bit more hospitable. Keep air circulating and the bees will regulate the interior climate to just the right temperature. Make sure the ventilation hole on the inner cover is turned upward and is open. For more details on providing additional ventilation, see Moisture Prevention on page 41.

→ Provide access to water. Bees need water to dilute honey and provide natural "air conditioning" to the hive. In early spring, natural water sources may still be frozen. Providing fresh, unfrozen water, situated in a sunny location, keeps the bees from looking elsewhere for a water source.

→ Remove swarm cells. If, during your inspections, you come across swarm cells, you'll need to decide what to do with them. Should increasing your apiary size be an attractive option to you, then you'll want to make a split, which is the noun used to describe "splitting" up one hive into additional hives (see page 76). Alternatively, you can simply remove and discard any swarm cells you come across. You'll need to be extremely fastidious in your removal, as leaving even one swarm cell behind may induce the hive to continue on with its plans to swarm. If you take them all out, the colony won't swarm; in the interest of self-perpetuation, it won't leave a colony without a queen to go form a new one. That said, once you find swarm cells, even if you completely remove them all, it's likely that the colony will make more pretty soon.

→ Make splits as needed. Splitting up colonies with swarm cells present on frames is one way to both prevent swarming and increase the amount of hives you have in your apiary. Alternatively you can make nucs to sell to other beekeepers.

Requeening

Sometimes the colony decides it wants to replace its queen. As discussed on page 71 in "A Look Inside," when the hive makes that decision (which is promoted by a change in pheromones produced by an aging or diseased queen), it begins creating supersedure cells. During your inspections, if you should discover supersedure cells (which are usually located on the upper portion of a frame, as opposed to swarm cells which are found on the bottom), you have two options: 1) *requeen*, meaning that you will physically remove and kill one queen and replace it with a new one that you purchase, or 2) allow natural supersedure to occur. There are strong advocates for both approaches. If the colony is allowed to supersede naturally, some time will be lost as the new, virgin queen must first mate with drones before she will begin laying eggs. That process takes about a month. Requeening is done with an already-mated queen, allowing for no interruption in egg-laying.

Personally, I prefer supersedure. I'm an overall "let nature take its course" sort of gal anyway, so this personal philosophy extends to my beekeeping practices. If I miss out on a bit of honey-production time, I'm not bothered. If I were in the business of making money from my honey, however, I might feel differently. In natural supersedure, the first virgin queen to emerge from her cell usually kills off any other queen cells right away. She may or may not then also kill her mother. Sometimes, both new and old queen will continue laying eggs until the elder queen dies or the hive kills her off (often by forming a ball around her, which kills the queen through heat exhaustion). On occasion, the first queen to emerge will swarm, taking half of the hive with her, when the other virgin queens emerge. This is called a cast swarm, and it can continue happening with each subsequent queen's birth. If that happens, unless you can recapture the swarm, you will eventually lose your entire colony. There is really no way to know which avenue the virgin queen will take: divide and conquer (killing off the competition) or moving on (throwing a cast swarm). Some beekeepers elect to remove all but one supersedure cell to prevent this from occurring. They do this only after discovering an opened queen cell, meaning the

virgin queen is present in the hive. Confident in the knowledge that at least one new queen has emerged and is alive, they will then remove any other supersedure cells.

There is some concern over limited genetic diversity in beekeepers' hives, prompting some to make the decision to requeen instead of allowing for supersedure. If many of the beekeepers in your area are ordering their bees from the same source, genetic diversity may become reduced. This could impair the queens' reproductive capabilities and the quality of eggs over time. However, if you know the bees in your area have been sourced from a variety of bee suppliers, this may not be an issue. The only way to know this, of course, is to get to know your fellow beekeepers. I cannot stress enough the importance of making the acquaintance of beekeepers in your area for reasons just such as this, in addition to the incredible wealth of information and assistance they may provide the novice beekeeper.

If you decide to requeen, the process is usually performed every two years. A queen is purchased from a respected queen supplier (and a local one, if at all possible) and introduced into the hive in her queen cage in the same manner as that performed when installing a package of bees (see page 59). On occasion, a colony will suddenly go *queenless*, meaning the queen perished without the colony having any time to prepare for it by building up swarm and supersedure cells. When this occurs, you can either purchase a new queen straightaway, or allow the hive to rear into a queen one of the youngest eggs the queen laid before perishing (this process is known as *emergency supersedure*). If the queen perishes with no young eggs present, however, then one of the workers will take over the responsibilities of laying. This is a no-win situation, however, as laying workers have never mated and will only produce drones. To attempt to prevent this situation from ever happening in the first place, move slowly and carefully while inspecting your hive so that you don't accidentally kill the queen and induce this panic situation. Staying on top of inspections and performing them regularly will also go far toward finding your colony queenless too far into the process.

SUMMER

As the heat of summer sets in, the hive is working at full steam. Colony numbers are at their peak, and nectar flows are abundant. All the bees are active, diligently toiling away at their individual roles, making all the necessary preparations for the colder months to come. The beekeeper's primary task in the warmest months is extraction. Other than that, the work tapers off during summer, unless a drought or heavy rains hinder the bees' access to nectar. In such cases, supplemental feeding will be necessary. Pay attention to weather forecasts in your area, check in on the hive periodically, make plans for extraction, and otherwise let the bees get on with the business of being bees.

Checklist

☐ Inspect the hive every week, monitoring the queen's activities, honey production, and presence of any swarm cells. Look for eggs, larvae, and other indications of active laying.

☐ Add supers as needed, for honey as well as for brood. Consider adding a second brood box if your hive seems to be "boiling over" with bees, an indication that the queen may need more room.

☐ Offer sugar syrup if your area experiences atypical weather, such as a drought, unseasonably cool temperatures, or excessive rainfall, as such situations may affect the bees' access to nectar sources.

☐ Install an entrance reducer if there is evidence of robbing or wasps or yellow jackets attempt to gain entry to the hive.

☐ Examine the varroa mite population and treat as desired (see page 87 for a discussion of varroa mites).

☐ Be prepared to extract, most likely in late summer (sometimes, an extraction in late spring or early summer is also possible). Have extracting equipment on hand, or reserve it in advance. Keep empty bottles and lids available. When you extract, be sure to leave adequate stores for the bees to get through winter (see page 99 for more information).

AUTUMN

As autumn arrives, the queen begins tapering off her egg-laying. Nectar flows and pollen sources begin to dry up while the still-busy bees fastidiously fill nooks and crannies securely with propolis, give the boot to any lingering drones, and start hunkering down for the winter. With a decrease in egg production comes a reduction in the colony's overall population. The preparations you make now will largely impact whether your hive makes it unscathed through the winter. Take the time to shore up your bees now, so that, come springtime, you'll all be basking in the sun's rays and reveling in the sprouting crocuses and apple blossoms.

Housed bees are much more likely to make it through the colder winter months than are wild colonies such as this one.

Checklist

☐ Check in on the hive every two weeks. Stop inspections completely once the temperature drops below 50 to 55°F (10 to 13°C) during the day.

☐ Confirm the presence of a queen. Look for eggs, capped and uncapped brood, and a good laying pattern.

☐ Assess capped honey stores. If they are low, you will need to feed supplementally (see page 99 for how to calculate how much honey is in a frame). Begin supplemental feeding in early autumn, around September, so that the bees will have enough time to convert your provided food to honey before the weather becomes too cold.

☐ Consider planting some winter-blooming bulbs such as snowdrops, hellebores, and crocuses to offer your colony a winter pollen source.

☐ Tilt the hive slightly forward to allow rain and snow to run off.

☐ Ventilate the inner cover (see page 41).

☐ Remove the queen excluder, if using.

☐ Stop all feeding by late autumn. Remove your hive-top feeder, if using, and replace the inner cover.

☐ Install a mouse guard.

☐ Install an entrance reducer, if not already using.

☐ Secure the hive with either a large, heavy rock atop the outer cover or a strap running the circumference of the hive (from side to side, not from entrance to rear).

☐ If you live in an area that experiences severe winters, insulate the hive by wrapping it with black roofing tar paper (see Wrap It Up on page 82 for more information).

☐ Store empty honey supers in a secure location, away from inclement weather, bears, and wax moths (see page 90).

WINTER

There won't be much for you, the beekeeper, to work on during the winter. That said, late winter and early spring are when the bees are at their most vulnerable. Hive losses occur more at  this time than at any other. Preparations made during autumn really come into play now. The bees will gather together inside in a football-shaped cluster to keep warm, periodically changing positions to access food stores. Total population numbers will have diminished considerably, balancing out somewhere around 10,000. Egg-laying, which tapers off in autumn, will resume again in mid to late winter.

Strong hives, offered plenty of their own food and given proper ventilation, should make it through winter unscathed.

Checklist

- Check the bee yard regularly for damage caused by storms or animals.

- Gently tap on the side of the hive. If you hear buzzing, the colony is alive. No need to open the hive, as you wouldn't want to expose them to the cold.

- Check that mouse guards are still in place.

- Check food stores by lifting the hive to determine weight (see page 44). Consider supplemental feeding if necessary (see page 45 for discussion of dry sugar and fondant feeding).

- If no unfrozen water source is available nearby (such as a creek or river), provide water; be sure to place it in a sunny location.

- Order more frames and supers, or other equipment, as needed.

- Continue reading and learning about honeybees.

Wrap It Up

In climates with severely harsh winters, where temperatures remain below freezing for months, wrapping hives with tar paper can be the difference between hives that perish and those that thrive. Tar paper, which is really quite inexpensive and readily available at most building supply stores, absorbs the sun's rays, warming the hive and melting any snow that may fall on it. As a consequence, the bees are able to break out of their wintertime cluster and move to honey stores. I've heard of bees dying of starvation with honey nearby, but just far enough away that they weren't able to break away from the cluster. Tar paper acts as a mini solar cooker, giving the hive just the right amount of warmth. Simply measure your hives and cut the tar paper with a craft knife. Wrap the tar paper around the supers and the outer cover, leaving the entrance open, and secure it in place with a stapler.

Body Count

If you should come to check on your hives one winter day and discover a number of dead bee bodies lying about, don't be alarmed. Casualties occurring during winter are normal, the natural passing of the hive's more senior members. House bees wait until a moderately warm and sunny day to remove the bodies. At other times, bees will leave the hive on winter days for "cleansing flights" (bathroom breaks, essentially), only to alight upon snowy ground, become chilled, and perish. This breaks my heart, but there is very little that can be done about it. The one consolation I find in it is that it is at least an indication that the hive is still alive, with its members strong enough to take flight.

Profile of a Beekeeper

Debra

A woman of multiple talents, career paths, and interests, Debra has enjoyed a love of bees throughout her life. Currently a natural beekeeping teacher, writer, and documentary filmmaker, she also can boast Master Beekeeper in her abundant apiary skill set. Debra has devoted her life toward honeybee advocacy, having founded The Honeybee Project to educate children about honeybees, as well as offering presentations on the mystical life of a beekeeper. Additionally, she is consulting with a variety of artists who are creating honeybee-influenced installations for museums and arboretums. Debra is also co-authoring a book about the sacred history of beekeeping, as well as another on her own "blessed journey" with honeybees. In her own words, it's clear that she is "bee-sotted."

Home for Debra is 15 acres in a rural setting in the mountains of western North Carolina. Attentive to the inherent benefits of a genetically diversified bee yard, Debra refers to her blend of Italians, Russians, various hygienic breeds, and introduced swarms as a sort of "European Union." Although she could accommodate more, she prefers to keep her apiary small, maxing out around five to seven hives at any given time, giving away splits or captured swarms to good homes. She does this because doing so helps her to stay in what she refers to as "intimate relationship" with each colony. Staying small, as she describes it, "helps me learn and be a better honeybee steward. I need to spend enough time with each hive to understand the singular entity that each hive is—what they sound like, what their preferences are, how to really work with them so they can flourish."

Moving into her fifth or sixth year of being "kept by the bees" (she claims she can't really remember on account of how "blissfully eternal" the relationship feels to her), this seasoned steward is rife with *Apis mellifera* wisdom. Her tidbits of advice are profound and heartfelt: "For anyone entertaining becoming a beekeeper, prepare to fall in love… and for generosity and gratefulness to well up and out of you so overwhelmingly that you feel like life is pouring a libation through you. Prepare to feel like dying of sadness the first time a hive crosses over because of ill health, and then let your heart break open. Prepare to have yourself reflected back to yourself every time you step into the bee yard… and to learn from that. Start finding a thousand ways to say thank you for all the blessings that being around bees will bring you…because thank you in itself will never feel adequate. And prepare to fall in love with yourself and all life more than you ever thought possible, for the bees will bring you home."

Chapter 9
Health & Wellness

Like all other living things, honeybees can get sick, they can get attacked, and they can get injured. They are just as fragile as humans, if not more so. Your job, as their steward, is to take action to keep them as healthy and happy as possible. If you're like me, you practice preventive care for yourself and your family and keep a vigilant watch for any sign of potential health problems, addressing them before they have the opportunity to advance into full-blown crises. You will need to exercise the same proactive approach with your bees. In this chapter we'll examine the most common diseases, parasites, and pests that threaten the health and well-being of your hives. Knowing what to look for during routine inspections will go far toward keeping trouble out of the hive while promoting health within.

DISEASES

American Foulbrood (AFB)

If "foulbrood" sounds awful, it's because it is. It's foul, vile, and ultimately, quite sad. Foulbrood, as its name suggests, is a disease attacking brood. Caused by the bacterium *Paenibacillus larvae larvae*, American Foulbrood is possibly the most devastating of all bee diseases. It is highly contagious, both within a hive and among neighboring hives. If AFB is detected, there is no remedy; the only treatment is to destroy the hive completely by fire. The entrance must be covered and the entire hive burned in a shallow pit, frames and all. The ashes should then be covered over by dirt so that no robbing can occur. AFB spores can live for many, many decades, so burning and burying infected hives is the only safe treatment solution.

AFB Characteristics

- Brood die capped
- Brood cappings sink inward and look punctured or perforated
- Brood pattern is spotty instead of compact
- Brood change color from a pearly white to a milk- to dark-chocolate brown
- Brood emits a sharp, sulfurous, foul stench
- Brood cells possess a "ropy," stringy, stretchy texture if punctured with a small stick

There are two antibiotics used to prevent AFB, Terramycin and Tylosin. Deciding on whether or not to use them is largely a matter of your overall approach to beekeeping. Like many beekeepers, I am opposed to routine antibiotic use, on the grounds that repeated applications of antibiotics eventually result in both antibiotic-resistant bacteria and toxic residues in honey and honeycomb. Instead, I advocate hygiene as the best form of prevention. If you have more than one hive, consider heat-sterilizing your hive tool between hives using a blowtorch, lighter, or lit match. Don't feed your bees any food from other hives that may possess AFB spores (if you use outside honey or pollen stores, confirm the supplier's hives are AFB-free). Don't purchase used equipment unless, as with food, the supplier is known and trustworthy. Lastly, if you are ever faced with the need to purchase a queen, consider hygienic queens. Reared with the genetic ability to both detect and remove perceived hygiene threats (including a number of bee diseases and hive pests), *hygienic* queens rear generations of bees possessed with the same self-maintaining qualities. These approaches would be considered Integrative Pest Management practices, or IPM. They exemplify the manner of beekeeping that I espouse and will be looked at more fully on page 88.

European Foulbrood (EFB)

Similar to American Foulbrood, European Foulbrood is a brood-attacking disease caused by the bacterium *Melissococcus plutonius*. EFB is harmful, but not nearly as devastating as AFB. European Foulbrood spores don't persist in the environment as AFB spores do; accordingly, if the disease manifests in your hives, such extreme measures as burning and burial are not necessary. Strong hives can often stave off the disease on their own, while weak hives may need your assistance. Frames containing uncapped, healthy larvae from strong colonies may be added to EFB-infected hives, allowing healthy larvae to attract attention from nurse bees to themselves, and away from diseased larvae.

Being able to distinguish between the two forms of foulbrood enables you, the beekeeper, to determine which course of action to take in handling the disease. I advocate the use of integrated pest management practices to prevent EFB.

EFB Characteristics

- Brood die uncapped, making them clearly visible to the beekeeper
- Brood pattern is spotty and random
- Brood cappings sink inward and look punctured or perforated
- Brood color changes from pearly white to an off-white or yellow-brown
- Brood often appear as though they have "melted" in their cells
- Brood emits a sour smell, but not nearly as bad as that produced by AFB
- Brood cells will not possess the "ropy," stringy, stretchy texture associated with AFB

Chalkbrood

Appearing most commonly during spring months when temperatures fluctuate between cool and warm extremes, Chalkbrood is another disease affecting brood. Caused by the fungus *Ascopsphaera apis*, Chalkbrood may occur when brood become chilled, allowing fungal spores to germinate and spread, moving into the larvae's gut and competing there with it for nutrients. The larvae may then die, permitting the fungus to invade and overtake the entire larval mass, rendering it into a white, hard, mummified body resembling a piece of chalk. The hardened masses are easy to remove from honeycomb cells, which undertaker bees do at their first fair-weather opportunity.

The first indication that your brood might have a Chalkbrood issue will be the appearance of their chalk-white bodies on the entrance board or on the ground in front of the hive. Chalkbrood isn't a devastating disease in the manner of AFB or EFB. For the most part, utilizing the IPM practices outlined for AFB will work to prevent Chalkbrood from ever appearing in your hives. Additionally, remember to consider the recommendations for hive siting suggested on page 40. Keeping hives out of damp, dank, low-lying locations, as well as providing good overall ventilation, will go far toward preventing Chalkbrood spores from finding a hospitable environment in the first place. Furthermore, some beekeepers suggest periodically culling old frames. Depending on their condition, completely replacing frames every two to five years eliminates any accumulated pathogens or toxins that may have built up on honeycomb cells. Finally, should your hive go queenless or you decide you'd like to requeen with purchased stock, consider a hygienic queen. Her offspring will be inclined toward culling any Chalkbrood-infested cells from the hive.

Sacbrood

Caused by a virus, this condition is considered somewhat rare. The Sacbrood virus infects brood during the prepupal stage. You will know it is present by the appearance of watery, limp, yellow-brown larvae. Sacbrood is somewhat analogous to the common cold. If supplied with good-quality food (like nectar and honey) and predisposed of a robust constitution, the hive should recover fully on its own. As suggested with Chalkbrood, culling and replacing old, worn frames with new frames periodically will also go far toward keeping Sacbrood in check. You may remove the Sacbrood-afflicted larvae from cells with tweezers or leave that task to the bees.

Nosema

Whereas the previously mentioned diseases are threats specific to brood, nosema is a condition affecting adult bees. Also known as dysentery, nosema is the most serious of all adult bee diseases. Caused by a single-celled protozoan, *Nosema apis*, the virus attacks the bees' digestive system. It most often appears in early spring, when bees have been cooped up due to cold or inclement weather and are subsequently unable to take "cleansing" flights to relieve themselves.

As worker bees go about the business of keeping the hive clean, they busily bite and lick surfaces within. In so doing, they may ingest *Nosema apis* spores.

Once consumed, the protozoan can seriously impair the bee's ability to digest food, resulting in reduced honey consumption and reduced colony buildup. In addition to a diminished post-winter colony buildup, visible characteristics of nosema include bees that appear to be disoriented, stumbling about the entrance of the hive. Also, nosema may evidence through streaking or spotting, evidenced by dark yellow-brown dots and streaks around the front of the hive and sometimes on frames. The disease may or may not kill bees. It will certainly weaken them.

Ways that you can prevent nosema include proper ventilation, siting the hive in sunny, dry, and draft-free locations, periodically culling frames (to remove spores that may live on in bee feces), and providing plenty of honey going into colder months. Some beekeepers treat preventively for nosema with the antibiotic fungicide fumagillin (Fumidil B) during autumn. As a beekeeper employing organic beekeeping practices with my hives, after careful reading and consideration on the topic, I have elected not to use this treatment. The decision to use or not use antibiotics with your bees is entirely up to you. Read up on the topic and make the choice that best suits your beekeeping approach.

PARASITES

While you're doing all you can to bolster and fortify your bees, a number of organisms, both large and small, are working in the opposite direction. All that the honeybee offers, from brood to pollen, honey, and a warm, cozy home, are terribly attractive to a wide variety of living things.

Varroa Mites

Perhaps the most serious condition affecting both bees and beekeepers alike, the ravages incurred by the *Varroa Destructor* mite have become an international problem. Tiny red or brown-colored parasites resembling miniature ticks, female varroa seal themselves away in brood cells just before they are capped. Once inside, they suck haemolymph (bee blood) from developing bees, resulting in deformities including chewed, deformed wings, transmission of viruses, and crippled, sickly bees that are essentially useless in the hive. Many larvae simply die in their cells, too crippled by varroa to emerge alive. As the queen's laying tapers off in late autumn and into winter, varroa populations decline, as there are fewer brood cells within which to feed and reproduce. Instead, the mite populations that remain huddle into the wintertime bee cluster, feeding off the haemolymph of adult bees.

If varroa mite infestation is left untreated, entire colonies may eventually disappear. The first line of treatment, then, is monitoring. By keeping a watchful eye over varroa populations, the parasite can be maintained at a level that will not harm the colony. Attemping to fight off infestation is often futile, as almost every beehive on the planet will have some level of varroa present. Complete removal of the parasites is less the goal than developing a manageable coexistence.

During routine inspections, you can detect the presence of varroa one of three ways: by seeing the mites on the bees themselves, by witnessing them on uncapped larvae, or by the presence of deformed wings on adult bees. If any of these signs are visible, employ one of the following methods to check varroa population numbers: powdered sugar shake,

opening drone brood cells with an uncapping fork, and utilizing sticky mats on screened bottom boards.

Powdered Sugar Shake

You'll need to gather up a quart-sized mason jar or other jar with screw-top threads, a canning screw ring to fit the jar's opening, and hardware cloth cut into the circumference of the opening and fitted into the screw band. You'll also need a sheet of plain white paper. Add $1/4$ cup of sifted, lump-free powdered sugar to the jar.

Don your protective gear, light your smoker, and open your hive, moving through supers until you reach the brood nest. Once there, scoop a large amount of bees (you're hoping to gather up around 300 bees) from off of the brood frames into your jar. Secure the jar lid and close the hive back up.

Place one hand over the screened lid and shake the jar repeatedly, until the sugar completely coats the bees. This doesn't injure them in the slightest, and their sisters will lick them clean once they return to the hive.

Remove your hand, and shake the sugar through the screening and onto the sheet of white paper. Uncap the jar and let the bees fly out (be sure you're wearing your protective gear, as they will be less than thrilled by your presence at this point).

Shake any remaining sugar onto the sheet of paper, and count the number of mites in the powdered sugar. If you count more than 10 mites, you've got a serious infestation on your hands. We'll discuss treatment options ahead.

Opening Drone Brood Cells

Varroa are especially fond of laying eggs inside drone cells. By checking a sample, you can assess the level of mite infestation in your hive. Here's how:

Don your protective gear, light your smoker, and open your hive, moving through supers until you reach the brood nest.

Integrated Pest Management

During inspections of varroa populations, if you should discover problematic infestation levels, you'll need to take action. The methods that I like to employ are collectively referred to as *Integrated Pest Management* or IPM. These strategies work to manage pest populations. They don't eradicate them completely, but ideally they keep intruders at a level that the hive can handle themselves. With IPM, regular inspections and prevention are preferred to acute treatment. Optimally, IPM approaches will reduce or omit entirely the use of hard chemical applications. So called "soft" treatments, which generally involve subtle hive and hive component manipulation, are employed with the goal of supporting the hive to naturally defend and heal itself from within.

Use the following techniques to support an IPM-based approach to managing varroa populations:

- Insert a screened bottom board so mites will fall through and populations may be monitored.

- Removal of drone brood cells

- Maintain strong colonies through routine inspections, providing proper ventilation, leaving the hive with proper food stores in winter, siting the hive in a sunny location, preventing robbing, and so on

- Purchase bees known for varroa resistance. Some bee suppliers advertise varroa-resistant bee strains, and hygienic queens are beneficial for removing infected larvae.

- Reduce robbing, as this can spread varroa between colonies

- Treat hives in spring and autumn with an essential oil-based supplement such as the much-touted *Honey B-Healthy*. Research indicates the product, while not killing mites, instead fortifies honeybee constitutions, thereby enabling them to fight off any parasitic attacks; treatment must be done when honey supers are not on, as the honey could take on the flavor of the essential oils; accordingly, it is usually administered before the spring honey flow and after the last autumn honey flow.

- Shake organic, sifted, cornstarch-free powdered sugar over the bees. Do this only if you are using a screened bottom board. Use either an empty powder bottle or any other sort of container with a sifter top, and shake a light dusting of the sugar over all of the bees on each frame. If this low-fi approach works as planned, mites will lose their footing on the honeybees, dropping down to the screened bottom board and out of the hive. Repeat weekly for the first few weeks of early spring, once the bees have broken cluster, and again in autumn.

Remove a brood frame and look for capped drone brood (these protrude out farther than their sisters).

Using either a bee brush or a firm shake, remove the bees on the frame (do this over the open brood nest in the event that the queen should be on the frame; you want her to go back into the brood nest and not fall onto the ground).

Move to the side or rear of the hive, turn your back to it, and, with the aid of an uncapping fork, open approximately a quarter to a third of a cluster of drone brood. Pierce the larvae and pull them out.

If you see more than two mites on a single larvae, you've got a serious infestation on your hands. Counting only a few mites in your sample indicates the presence of a small varroa population.

Sticky Screened Bottom Boards

In order to utilize this method, you will need to have screened bottom boards in place. Screened bottom boards are a good idea in any case; under normal circumstances mites fall through, leaving the hive, never to return. With a solid bottom board, mites that fall to the bottom of the hive simply hitch a ride on the next adult bee to fly by.

Most of the year, the screened bottom board is open, permitting optimal ventilation inside the hive. During inspection times, the removable plastic insert that comes

with screened bottom boards is inserted. Before you insert the plastic board, coat it with a sticky substance such as vegetable oil. Now, mites that fall off of bees will collect on the bottom board and be stuck.

After three days, remove the bottom board and count the number of mites gathered. Divide the total number by three, which will give you a rough daily mite fall number. Between 20 and 50 mites total is tolerable. If you count more than 50 or so mites, you have an infestation and will need to do something about it.

Tracheal Mites

Considerably smaller than their varroa cousins, tracheal mites are undetectable to the naked eye. Accordingly, their presence cannot be monitored via inspections. As their name indicates, these pesky creatures live in the thoracic trachea of honeybees, where they reproduce. Tracheal mites bite into the tracheal (breathing) tube and suck on the bee's haemolymph. Eventually, tracheal mites so fully congest the breathing tube that the bee is unable to get enough oxygen to fly.

Absent a microscope for performing thoracic inspections of deceased bees, it's almost impossible to detect the presence of tracheal mites. Subtle clues can give you a heads-up that there might be a problem, however. If you notice bees stumbling on the ground in front of the hive or trying unsuccessfully to fly, you might have a tracheal issue. Grease patties infused with essential oils can help prevent or treat tracheal mite infestation. Once consumed by honeybees, the oils in the patties cause the resident mites to become confused in their search for a host. Grease patties have the added benefit of reducing varroa mites, as the grease prevents mites from securing a strong hold on the bees. Accordingly, they fall to the bottom where, if screen bottomed boards are in place, they will then fall completely out of the hive.

Grease Patties

1. Mix 2 parts granulated sugar to 1 part saturated fat, such as all-natural vegetable shortening or coconut oil. The exact amount doesn't matter so long as you adhere to the 2:1 sugar-to-fat ratio.

2. To this mixture, add ¼ to ½ cup honey as needed to make a soft-but-not-runny consistency. Stir 5 to 10 drops of wintergreen, spearmint, or peppermint pure essential oil into the mixture. The finished product should be quite stiff.

3. Using an ice cream scoop, scoop uniformly sized lumps and press the mixture into patties with your hands. Repeat until all of the mixture has been formed into flattened patties.

4. Place one patty directly onto the top bars of the brood chamber. Allow the bees to consume the grease, and replace with a new patty as needed. Store additional patties between sheets of wax paper in a sealable bag or lidded container in the freezer until needed.

PESTS

Small Hive Beetle (SHB)

Technically known as *Aethina tumida*, the small hive beetle was introduced to European honeybees via Africa. Easily handled in their native environment, this exotic insect has the potential to wreak havoc on small, weak colonies. Dark brown to black in appearance once mature and slightly larger than a tick, female small hive beetles lay eggs inside the hive. Their larvae go on to eat pretty much everything inside the hive except for the honeybees themselves—pollen, larvae, honey, wax, eggs, the works. As if that weren't bad enough, they then leave an oozy trail of excrement, which, should it pollute uncapped honey stores, causes fermentation and a subsequent bubbling mess in the hive. Small hive beetle larvae slowly work their way down and out of the hive, heading toward soil where they will burrow and pupate, "sliming" the entire hive on their way out.

As a tropical pest, small hive beetles are of smaller concern to beekeepers in northerly climates, although northern hives should not be considered immune. The best measure of defense against small hive beetles is to keep colonies strong by following IPM practices such as providing appropriate ventilation, adequate food stores, and proper hive siting, and using genetically resistant bee stock. Consider the use of beetle traps if you have a problem.

The most readily available models of small hive beetle traps are the Hood trap and the West trap. The Hood trap is designed to capture adult hive beetles and consists of a small plastic tank or reservoir that is attached to the interior bottom edge of a frame. The tank is filled with a mixture of vinegar and oil to attract and trap the beetles. The West trap is designed to trap larvae on their exit route. Attached to the bottom board, this model of trap contains a tank running the length of the hive (much like a hive-top feeder). The tank is filled with oil, into which the larvae will fall and become trapped. A grip running the entire length of the tank prevents bees from falling in as well. Both types of small hive beetle traps can be sourced from many beekeeping suppliers.

Wax Moths

The greater wax moth, *Galleria mellonella*, is often merely a nuisance. If allowed to flourish, however, this scavenger can destroy a hive. Once female wax moths have penetrated the hive, they lay eggs in the brood box. Their young then tunnel through brood comb, leaving silvery, threadlike webbing and feces everywhere. During warmer weather, they can ravage entire frames of comb in as short as two weeks. Strong hives will often manage wax moths themselves, while small or stressed hives are more susceptible to infiltration.

The best line of defense against wax moths is to practice proper storage of empty honey supers (refer to Cold Storage on page 109 for storage suggestions). Also, you want to be sure to add honey supers only when the supers below are almost completely filled. Doing so minimizes the spaces in the hive that are absent of bees; sufficient, strong numbers of bees in a hive can patrol, defend, and ward off interlopers in empty cells. After extraction, replace the extracted supers back onto the hive, taking them back off again after the first hard frost; this allows bees time to remove any residual honey.

It's possible to also practice preventive treatment against wax moths (however, this method won't work after an infestation is present). An annual spray of diluted Bt (*Bacillus thuringiensis*) over the wax comb on each frame has proven almost 100 percent effective at deterring wax moths. Bt is a so-called "soft" treatment consistent with integrated pest management and is considered an alternative to paradichlorobenzene, a synthetic, aromatic compound similar to that found in moth balls. This chemical is not permitted in organic beekeeping practices.

Ants

More of a nuisance than a serious problem, a few ants in the hive are fine, but too many can make a hive head for more hospitable environs, especially one that is small or weak. To keep ants from becoming a major issue, elevate your hives on a wooden hive stand or cinder blocks. If your hive is situated in an area where grass grows, remain vigilant about keeping the grass trimmed in the bee yard.

If ants are moving into the hive en masse, put the hive stand's legs in containers of water or oil. The water or oil moat will prevent ants from gaining entry by creating a rather insurmountable barricade.

Mice

Come cooler weather, mice will start searching for warm, toasty housing in which they can overwinter. You don't want to give them any opportunity to decide that your hives will make ideal lodging. If allowed entry (which they will do at night, as the colony begins to form its cold-weather cluster), they will eat pollen, chew through the woodenware, bring in grass and leaves, and generally make a mess of the place. Mice aren't a problem during warmer weather, as an active hive will thwart their entry with multiple stinging attacks. Once the bees are clustered, though, the mice can sail past the entrance unnoticed. While they

pose no direct threat to the bees themselves, they can destroy precious food stores and equipment.

To stop a mouse or any other rodent (you definitely don't want rats in your hive, *ever*) problem from developing, install mouse guards over the hive's entrance when the weather begins to cool in late summer or early autumn. Galvanized strips of metal (usually zinc) perforated with multiple bee-sized holes, mouse guards will keep mice out while allowing both air and bees to circulate within. Remove the mouse guard in the spring when the cluster begins to break and activity inside the hive resumes.

Skunks, Raccoons, and Opossums

For the most part mere pests, these nocturnal foragers can cripple weak colonies, as well as strong colonies left open to repeated assault. A skunk's calling card will be the presence of scratch marks at the hive's entrance. Guard bees come out to check on the disturbance, only to be gobbled up (skunks seem to be immune to stings on their paws, mouth, or throat). They'll repeat this process a number of times, returning the next night for another free meal. If you detect scratch marks or muddy paw prints on the hive entrance or landing board, or torn up sod or mulch on the ground in front of the hive, you might have a skunk problem on your hands. Opossums operate in a similar manner, only minus the presence of scratch marks. Another surefire sign that your hives are under nighttime attack will be an aggressive disposition. Agitated, angry bees can be a hallmark of a hive on extreme alert.

If you sense you might have a skunk or opossum issue, there are three lines of defense you can pursue. First, elevate your hives off the ground. Skunks or opossums would then have to expose their vulnerable (and, thankfully, sensitive) underbellies in order to access the bees. The stings will turn them off of their pursuit in short order. Secondly, you could hammer tacks or nails through a piece of plywood and situate the board on the ground in front of the hive. Just don't forget that it's there when you approach the hive, or you'll get the rude awakening intended for your thieving intruders! Lastly, you can cordon off the area in front of the hive by either placing a large roll of metal fencing in front of the entrance, secured down with a brick, cement block, or heavy rock, or stake metal rods into the ground and thread chicken wire fencing around them as a protective barricade.

Raccoons are less interested in the bees themselves, preferring instead the bees' pollen, honey, propolis, and wax. Marauding raccoons will remove the outer and inner covers of a hive and pull out frames. They'll then take the frames over to a corner of the bee yard and greedily munch away. Securing the lid of the hive with a heavy rock or brick can easily prevent this entire debacle. Because the scent of wax or other bee products can serve as an attractant to raccoons, avoid discarding any hive by-products onto the ground of your bee yard. When performing hive inspections, gather up any debris in a lidded jar and remove it from the area.

Bears

Honeybees have a long, storied, and rather controversial, history with bears. Bears love to eat honey and brood—it's almost as simple as that. Bears are also big and strong, and not particularly bothered by bee stings. Accordingly, if allowed entry to your bee yard (often at night), they will not only decimate almost every hive in sight, they will remember their tasty exploits, returning again and again. As such, the best defense is advance precautions.

If you live in an area known to possess bears (as I do), make every attempt to keep your bee yard free of any

equipment or hive debris, including empty supers, wax, drone cells, and so on. For greater security, consider installing electric fencing. In order for electric bear fencing to be effective, it needs to be well grounded, properly charged at all times, and the area around it needs to be regularly maintained (grass and weeds need to be trimmed, branches must be kept off the fence, etc.). Don't choose a location adjacent to trees, as branches may fall and disarm the charge; determined bears might also climb the tree, dropping down directly into the bee yard. The voltage will need to be checked periodically with a handheld portable device called a voltmeter. Fences can be either permanent or temporary, but should consist of wires 8 to 10 inches (20.3 to 25.4 cm) apart. The bottom wire should be 8 inches (20.3 cm) from the ground, and the fence itself needn't be any higher than 3 to 4 feet (.9 to 1.2 m). Batteries for electric fencing may be either electric or solar and need to support at least 12 volts. Whichever option you select, the battery must be reliable and continually charged.

Beekeepers refer to the heartbreaking act of a bear destroying one's hives as having been "Yogi-ed," a reference to the cartoon character known for his robust appetite for honey. Should you suffer the devastating ravages of a bear assault, compensation for your losses may be possible. If you live in an area determined to be a known bear habitat, register your bees with your local wildlife or conservation agency. "Bear" in mind, however, this will need to be done well in advance of an actual attack.

A bee yard utilizing bear fencing

Close Encounters of the *Ursidae* Kind

The following is an excerpt from my blog, Small Measure (www.small-measure.blogspot.com), relaying an up-close and personal interaction I had with a bear one hot, steamy August afternoon. Prior to this sighting, neither my neighbors nor I were aware that we had bears in the vicinity:

I have a visitor. While it should come as no real surprise, given that I live way at the end of a dirt road, in a forest, surrounded by a 350-acre nature preserve of yet more forest, I have to admit that I wasn't quite prepared for my visitor. To begin, I wasn't properly dressed, which is so important when greeting new guests, I've always maintained. Specifically, I was wearing my bee suit. I was also wearing knee-high rubber boots.

When my visitor arrived (we'll call her "Maude," or, on the chance that this might be a male guest, "Seymour"), I had just finished working in my beehives. I was enjoying the late afternoon summer breeze and waning sunlight. Suddenly, my dogs were off, in hot pursuit of something, dashing up the mountain adjacent to my bee yard. I lumbered after them, rappelling myself up a steep embankment, grabbing onto skinny, slippery tree branches for support (still in my bee suit, plus gloves and boots, mind you).

Once I reached the top of the embankment, I saw them chasing a black, rather large creature back down the mountain and directly toward me. I first thought, "Oh, they've found another dog to play with." This quickly morphed to "Oh, that's a wild boar," immediately followed by a stunned (and expletive-replete, I might add), "Oh, that's a bear!" And so, against all the advice I have ever heard about proper bear-greeting etiquette (make yourself look large, make a good deal of noise, be conspicuous), I fled, in holy terror. Fled is actually an inappropriate descriptor. Trudged, plodded, and otherwise clumsily, heavily ran are much more fitting descriptions of my terror-induced attempt to hightail it back down the mountain and then up the hill back to my house. Like I said, I wasn't properly dressed for receiving company.

The long and short of it is that, yes, we have a bear. Where there are bees, and corn (our land-share farmer has corn planted in the field below our house), and forests, it is entirely likely that there will be bears. Black bears have a reputation of being really rather benign, as far as bear dispositions go. Nonetheless, its presence necessitated a trip to the Tractor Supply store to pick up all the fixings needed for electric fencing. Our friend, the mighty of strength and generous of spirit Lance Graves, came over that evening and tricked out our new fortress in no time. On his way out, he said the bear crossed the driveway, and, to his experienced eye, probably weighed around 250 pounds.

The lesson learned from this experience is that urban (and especially rooftop) beekeepers have it made. Also, an ounce of prevention (and some common sense about living in an environment that bears share) is worth a pound of honey. Oh, and don't try running in your Wellies; it just brings all those dreams to life where you're trying to run and can only creep along frighteningly.

Colony Collapse Disorder (CCD)

It's not just honey-obsessed bears or opportunistic diseases that pose risks to our winged friends. Entire populations of honeybee colonies have begun disappearing completely. Poof. Gone. It would be all good and well to joke about alien abductions or other fantastical notions were the issue not of such grave concern. In 2006, beekeepers in the United States, upon performing routine inspections, began finding a good number of their hives to be totally empty of bees. Losses began to mount, with beekeepers reporting greater and greater numbers of completely empty beehives. Dubbed Colony Collapse Disorder, the mysterious syndrome has since spread, appearing internationally and affecting commercial and hobbyist beekeepers alike.

The jury is still out on what is behind CCD. It may stem from a single source or could very well be multi-causal. An international cadre of scientists is working with feverish dedication toward identifying the culprit. No consensus currently exists, and no definitive causes have yet been determined. Some of the more widely espoused potential causes include:

→ Varroa mites

→ Nosema

→ Environmental change-related stresses such as drought and climate change

→ Malnutrition, possibly from securing nectar exclusively from one source (more common in commercial-scale beekeeping)

→ Pesticides

→ Insecticides such as neonicotinoids

→ Migratory beekeeping and the stress it could be causing honeybees

→ Cell phone radiation

→ Genetically modified (GM) crops

What then characterizes Colony Collapse Disorder? While there is some variation between hives, CCD-affected colonies largely evidence the following traits:

→ A complete absence of bees, with no or little dead bee buildup near the hives

→ The presence of capped brood

→ The presence of honey and pollen food stores

→ The queen is present

→ The overall workforce at the time of collapse consists largely of young bees

→ The workforce is reduced in overall numbers

→ Other potential hive robbers, such as wasps or other honeybees, avoid the hive, even though it is unguarded and full of food

The greatest threat posed by CCD to humans concerns our food supply. Honeybees are responsible for the pollination of over 30 percent of the foods that many people regularly consume. As we all have to eat to survive, the threat holds potentially devastating consequences. Furthermore, many livestock animals rely on honeybees to pollinate the crops they consume. Absent honeybees, many crops would require hand pollination, a highly labor intensive endeavor. The diligent work of these pollinators saves an enormous amount of money for the producers of agricultural products, averaging somewhere around $15 billion annually. Other native pollinators such as wasps, bumblebees, and butterflies cannot be managed or transported in the same manner as *Apis mellifera*. While an identifiable cause of CCD and corresponding cure are being researched, many agriculturists are making efforts to encourage alternative native pollinators to their crops. Until we know more about its causes, I feel a beekeeper's best defense against CCD is practicing integrated pest management techniques.

A BEE GARDEN

An indispensable way to promote the well-being of honeybees—both wild populations and those in your care—is to provide them with abundant sources of nectar and pollen. Four seasons' worth of offerings will help keep our winged beauties in food supplies year-round, creating safe and nurturing habitats. The end result is that we get to enjoy expertly pollinated, fragrant, eye-pleasing, and delicious plants while the honeybees delight in readily accessible food stores.

When selecting plants and trees for both you and the bees, consider heirloom varieties, which are more likely to contain the greatest quantities of pollen and nectar. I'd also highly encourage the use of organic, pesticide and insecticide-free gardening practices. If you're new to those techniques and uncertain how to proceed, a great number of books are available on the topic. Check around your area to find organic gardening and IPM classes.

The following list highlights a number of flowering trees and plants known to attract honeybees. Though far from exhaustive, it presents a broad base of varieties and offers something for every season. Feed your bees throughout the year and they'll work valiantly toward feeding you in return!

FLOWERING TREES

- → Apple
- → Basswood
- → Cherry
- → Linden
- → Maple
- → Orange
- → Pear
- → Plum
- → Sourwood
- → Tulip poplar
- → Willow

FLOWERING PLANTS

- → Angelica
- → Anise hyssop
- → Aster
- → Basil
- → Bee balm
- → Black choosy
- → Borage
- → Butterfly Bush
- → Calendula
- → Catmint
- → Catnip

- → Clover
- → Coneflower
- → Crocus
- → Daffodil
- → Daisy
- → Dandelion
- → Flowering quince
- → Goldenrod
- → Heather
- → Honeysuckle
- → Hyacinth

- → Lavender
- → Lemon balm
- → Marigold
- → Meadowsweet
- → Mullein
- → Nasturtium
- → Nettle
- → Peppermint
- → Poppy
- → Rosemary
- → Sage

- → Snowdrop
- → Spearmint
- → Stonecrop
- → Sunflowers
- → Sweet pea
- → Thyme
- → Winter aconite
- → Wintergreen
- → Witch hazel

Chapter 10
Honey

A popular reason for getting into beekeeping in the first place, honey is the sweetest reward for your labor of love. Locally sourced, backyard (or rooftop!) collected honey far surpasses its processed, grocery store kin in both flavor and overall quality. Modern beekeeping practices result in the production of a surplus of honey, benefiting bee and beekeeper alike. Here we'll examine all the crucial details relating to how the liquid gold is created, extracted, stored, and sold.

MAKING MAGIC

From plant to capped comb, the manufacture of honey is a multi-step, labor-intensive, fascinating process. The sweet substance we, and the bees, crave begins as a plant secretion known as nectar. Found on a wide range of vegetation, these tiny beads of sugary liquid consist of mostly water (around 80 percent) and several complex sugars. As described in chapter 4 (see page 43), as honeybees gather up nectar from plants, they also gather pollen, transferring it across the plant's reproductive organs and enabling pollination in the process. The plants get a helping hand in propagating themselves while the bees get a food source. Everyone wins.

To access nectar, a honeybee uses its skinny, hairy, tube-like tongue, known as the proboscis. While foraging, the bee stores the lapped-up nectar in its honey stomach for safekeeping until it returns to the hive. Honeybees have two stomachs, one of which is expressly dedicated to holding nectar. This honey stomach holds almost 70 milligrams of nectar and, when full, will weigh almost as much as the bee herself. In order to gather enough nectar to fill its honey stomach, a bee will visit anywhere from 150 to 1500 flowering plants daily.

Foraging bees generally travel up to 2 miles (3.2 km) to gather nectar. If nectar offerings in that range are paltry, they will occasionally fly as far as 5 miles (8 km) to secure a source. The nectar not only provides the raw material for what will become honey, but it also provides fuel for the long-haul flights foragers make searching for it and food for growing larvae.

Upon returning to the hive, the foraging bee will regurgitate the contents of its honey stomach. A house bee then puts a drop of nectar on its proboscis and sort of "chews" it over for 20 to 30 minutes. As the bee chews on the nectar, exposing it to air, its digestive enzymes begin converting the nectar from a complex to a simple sugar. This step reduces the water content in the nectar, which in turn works toward preserving the honey against opportunistic, moisture-loving bacteria once it is stored. After the house bee has chewed on the nectar, it deposits each tiny bead into a honeycomb cell, repeating the process until a cell is filled. The nectar will still have a rather high proportion of water at this point; if left in its present state this moisture could induce the inherent, wild yeasts within it to ferment the simple sugars. Bees intuitively know this and get busy fanning the stored nectar furiously with their wings to further reduce the moisture content. As the moisture in the nectar diminishes, the sugars become more concentrated and thicken, thereby inhibiting fermentation.

Honeybees will continue fanning the stored nectar until, mysteriously (and, in my humble estimation, miraculously), they determine it has reached a moisture content right around 18 percent. Once this occurs, they will then cap the cell over with a thin layer of beeswax. The honey will be stored therein for use as a winter and early spring food supply. Depending on its size, a hive can produce between 120 and 250 pounds (54 and 114 kg) of honey annually.

Beekeeping practices, which utilize supers or other vessels to house colonies and supplement feeding when necessary, encourage a surplus production of bee products. A greater amount of honey is stored in a hive maintained by a beekeeper than would ever be found in a wild colony. Accordingly, beekeepers can safely take off excess honey stores without endangering the hive. Provided beekeepers are mindful of leaving enough honey for the hive to survive on during colder months when nectar sources are scarce, taking off honey from the hive doesn't really "rob" them, as honey removal is sometimes referred to.

THE HONEY HOUSE

You may have noticed—honey's sticky. Really sticky. Stick-to-every-surface sticky. It's also, unsurprisingly enough, really attractive to bees. As such, you should plan your extraction location mindfully, creating your "honey house" with care and attention. By definition, a honey house is simply a building used for extracting honey and storing beekeeping and extracting equipment. While I love the idea of a building devoted exclusively to the bees and all of their needs, I'm going to guess that many of you, like myself, lack the luxury (or space) of fashioning such a dedicated structure. Unless you have a business based around honey or other bee products (beeswax, propolis, pollen), more than likely you'll be extracting only once or twice a year. In that case, any indoor area such as a garage, basement, outbuilding, barn, or kitchen will suffice as an occasional honey house.

Wherever you choose to extract, the single most important concern is that it can be fully sealed off from bees or other curious, honey-loving, winged creatures, like wasps. Not only is there the risk of being stung (less from the honeybees than the wasps, as bees are generally disinclined to sting when away from the hive), but also, if permitted to gain access to the honey, the bees will make short order of carting it drop-by-drop back to the hive. So, find a sealed, fully enclosed area to work in. If at all possible, stage your extraction in an area with a floor that can be mopped. Either way, lay down a good amount of newspaper and have extra on hand for inevitable spills or messes. A water source is also extremely handy. Otherwise, keep several buckets of warm water on hand for rinsing off hands, knives, and other equipment. This is also not the time to have your dog or cat wandering around, full of curiosity, not to mention loose hairs. We all love our pets, but ultimately, no one loves finding animal hair in their honey.

THE HARVEST

Honey harvest time is one where a sticky situation offers a much-loved reward. Let's examine what's involved in knowing when, as well as how, to take off honey, and beeswax, from your hives.

Timing Is Key

Knowing when to harvest is key. If you remove honey too soon, you run the risk of collecting honey that still possesses a high degree of moisture, which could, in turn, cause fermentation once bottled. On the other hand, if you harvest too late, you might be dealing with honey that the bees have already begun eating or honey that is too firm and solid in the honeycomb cells to extract. Late removal of honey supers can also incite robbing. The perfect time to harvest is basically when the bees give you the go-ahead to do so. That signal is indicated by the presence of white, sealed or "capped" honeycomb covering 90 to 95 percent of the frame. For the most part, honey that is not yet capped contains a moisture level greater than that pivotal 18 percent. When the bees cap over the honey, they are telling you that the moisture level is just where it needs to be to prevent the natural yeasts in the nectar

from fermenting over time. Fermented honey isn't really palatable by either humans or honeybees, so bide your time, let the bees do their thing, and extract when the proper time arrives.

In addition to determining when the honey is ready, you'll want to make sure your calendar is clear come extraction time. Depending on the method of removal you use, it's possible for the entire process to take several days. Some options, such as bee escapes (see page 102), require 36 to 48 hours to work. After you clear the supers of bees, you'll then need to schlep them to the extraction site (which should be all ready and primed for action in advance), extract the honey, bottle it, and then clean up the whole sticky mess. I say give yourself a weekend, or some similar 2- to 3-day stretch of time. You don't want to rush this, nor can you, really. Good things (of the sweet, sticky, oozy, delicious kind) come to those who wait.

Some for You, Some for Me

In addition to knowing *when* to extract, it's of equal importance, if not more, to be aware of just *how much* you can extract. The bees make honey for themselves.

Yes, human intervention in the form of modern-day beekeeping permits them to make a surplus. However, they still need some if they are to survive during colder months. It is necessary to leave the bees 60 to 70 pounds (27 to 32 kg) of honey for hives situated in colder climates, and 20 to 30 pounds (9 to 14 kg) for those in warmer locales. Two methods of assessing honey stores are described on page 44. When full of honey, a shallow frame should weigh roughly 3 pounds (1.4 kg), a medium frame 4 pounds (1.8 kg), and a deep frame 6 to 7 pounds (2.7 to 3.2 kg). So, let's say your honey is stored in shallow supers. Let's also say that you have 10 frames full of honey in those shallow supers. Finally, we'll add that you live in a location with cold winters. In order ensure that your bees will have adequate stores to get them through the winter, you'll need to be certain to have at least two full shallow supers (totaling 20 frames of honey) on top of the hive body. Or, say you have a medium super, also with 10 frames full of honey. That's 40 pounds (18 kg) of honey, meaning the bees will need another super with 20 to 30 pounds (9 to 14 kg) of honey in it to get through the winter. Make sense? Anything after that is gravy and is perfectly fine for you to extract.

A Helping Hand

Although it is entirely possible to extract honey with one person, an extra set of hands is invaluable. If you have a good number of supers you'll be extracting, having another person around to help you haul them to the extracting site will streamline the entire process. It also expedites things to have one person uncap frames while another works the extractor. Finally, clean up and bottling is always easier with four hands instead of two. Come extraction time, ask around. You'd be surprised at just how many people are curious about honey extraction, especially if there's an of-fer of a jar or two of honey in exchange for their labors. A young child or even a non-beekeeping buddy may be will-ing to pitch in and process the liquid gold.

Fully Equipped

Several tools are necessary in order to extract honey. Depending on the size of your operation, you'll need to decide whether or not it's worth the expense of purchasing your own equipment. For many, extraction is an endeavor done once, possibly twice, annually. If you are a hobbyist beekeeper, with fewer than 10 or so hives, it might not be worth purchasing equipment solely for personal use. Many beekeeping organizations rent out extracting gear for a very nominal fee (as mentioned previously, mine rents for a really quite reasonable fee). If you don't have a beekeeping organization in your area, but find the cost of purchasing equipment prohibitive, try to connect with other area beekeepers and see if they might be willing to loan it out to you. Offer to pay, cook them a nice dinner, babysit, pet sit, or some other skill you possess in exchange.

Uncapping Knife

Essential to your first step in extraction, an uncapping knife removes the wax caps on honeycomb cells. An electric knife provides its own heat, much like an iron. If you don't have an electric model, a simple serrated bread knife will do the trick, as long as you dip it in hot water between uses (be sure to thoroughly wipe off the water before touching the comb, though, as you don't want to introduce any outside moisture to the honey).

Uncapping Comb

Sometimes uncapping knives miss several, or many, sealed cells. You don't want to miss out on any of the honey tucked into the honeycomb. That's where an uncapping comb, also known as a scratcher or fork, comes into play. Very much resembling a long-tined hair pick, the uncapping comb is used to rake, or scratch, down the length of the frame, releasing any trapped honey. Some beekeepers prefer the uncapping comb to the knife and use it exclusively. Try both until you can determine what works best for you.

Wax Cappings Basin

As you scrape off the wax seals, you'll need somewhere to keep them. They contain an awful lot of honey, so using a wax cappings basin with a built-in strainer is ideal. The wax cappings fall off the comb as you cut them and into a tub lined with a strainer, grate, grid, colander, or similarly perforated panel. While they rest there, the honey on them drips down, collecting in the lower basin. Ideally, this tub will have a spigot, allowing for easy removal of the honey collected at the bottom of the tub when you're finished uncapping frames.

Extractor

A device using centrifugal force to draw honey out of uncapped comb, an extractor is the best, and fastest, means of honey removal. There are as many models of extractors as there are types of beekeepers. Inexpensive, albeit less durable, extractors are usually made of plastic and are manually operated (meaning you, with your arm muscles, will be providing the centrifugal force). More rugged outfits are made of stainless steel, available with or without electric motors. Again, assess your needs, finances (the more powerful, mechanized models can cost hundreds of dollars), and resources communally available, and select the extractor that best suits your purposes.

Bottling Bucket/Dispenser

During extraction, honey will flow out of a spigot on the extractor and into a collection vessel. Ideally, this container will have a spigot itself, so that, come bottling time, dispensing honey into jars can be handled with ease. Your collection vessel should be either food-grade plastic or stainless steel. Available for purchase from beekeeping suppliers, 5-gallon (19 L) lidded bottling buckets will hold around 60 pounds (27 kg) of honey. After extraction, be sure to place a lid over your collected honey. Its inherent hygroscopic properties can pull ambient moisture from the atmosphere, possibly resulting in fermentation.

Filter/Strainer

Before you can bottle up your sticky sweet abundance, you'll need to filter it, removing any extraneous debris. Beeswax, bits of wood from the frames, even, sadly, bee parts can all potentially wind up in the honey. A fine mesh filter does this job expertly. Available from beekeeping suppliers, filters intended expressly for honey fit over the top of honey bottling buckets. It's also possible to use other sorts of meshed sieves, provided they are large enough (plastic paint strainers available from paint stores work well). Honey flows pretty fast when it's coming out of the bottling dispenser. Use too small a filter and you may end up with a sticky mess of honey overflow.

from head to toe. You'll also need a bee-tight receptacle to store the frames or supers as you collect them. If you're extracting from just a few frames, large, plastic tubs or containers with tight-fitting lids will work. If you're removing full supers, you'll need bee-tight covers on both the top and bottom of the super. Use either two pieces of plywood, plywood on the bottom and a heavy cloth on top, or painter's plastic above and below. Whatever you choose, just be sure it's sturdy enough to securely keep bees out.

Shaking or Brushing

A bee brush or your gloved hands are all you need for this low-tech means of removing bees from the hive. Hold a frame by the ends of the top bar, and then gently shake or brush bees off of each individual frame in front of the hive entrance. Then place the cleared frames in an empty, waiting super, and cover with a bee-tight top and bottom. Repeat the process until you've gone through each frame bound for extraction. This means of removal works fine if you've only got a hive or two. Bigger bee yards necessitate other evacuation plans, as shaking or brushing is a time-consuming endeavor. If you opt to remove bees using a bee brush, be sure to start at the bottom of each frame and move upward. Stroking the brush in a downward motion can injure or even kill bees working inside cells due to the inherent downward slant of honeycomb cells.

Escape Plan

When harvesting time comes, you'll need to remove the bees from the supers you intend to extract. You've got a number of different evacuation options. Read them over and see what seems like the best fit for your situation. Ask fellow beekeepers their preferred method, as well—most will be all too willing to opine. Whichever you choose, fire up your smoker first and give the hive a gentle smoking (not too much, or you may affect the flavor of the honey). The bees aren't especially pleased when honey is removed, so wear protective gear, too,

If removing bees with a brush, be sure to brush upward.

Blowing

Using either a specialized bee blower (available for purchase from beekeeping suppliers) or a basic backyard leaf blower, this means of removal essentially blasts the bees out of the hive. Remove the supers intended for extraction, and move them, bees and all, about 20 feet (6 m) from the entrance to the hive. Place the supers sideways on or directly beside whatever you intend to use for transport, be that a car, wheelbarrow, wagon, or what have you. Then engage the blower and point it directly toward the supers, where it will forcibly blow the bees out and away. Bees don't like this, mind you. Should our situations find themselves reversed, I don't imagine I'd like being blasted away from my home either. That said, blowing as a means of removal can be highly efficacious if you've got a load of supers to attend to. Remember to cover the supers with bee-proof tops and bottoms once cleared of bees, and then place them in whatever you'll be transporting them in straightaway.

Bee Escape

Another means of clearing bees out of supers bound for extraction, bee escapes are devices added to the hive. They allow bees to head down and out of the hive, but not back in the same way, at least, not right away. It takes about 36 to 48 hours for bee escapes to successfully vacate a super. However, it takes right around that same time frame for bees, clever creatures that they are, to figure out how the escape works and return. So, if you intend to use a bee escape, know in advance that you'll be making two trips out to your bee yard—first to install the bee escape, and then another, a few days later, to remove it.

Though beekeepers have utilized bee escapes of varying styles for years, two models are the most frequently used in modern beekeeping: the Porter bee escape and the triangular escape. Porter escapes, made of either plastic or metal, are placed over the inner cover opening. The inner cover, with the Porter escape in place, is then situated between the supers bound for extraction and those that will remain in place. Triangular escapes consist of a wooden cover with two, interlocking sets of triangles in the center, which are covered over with wire mesh. Like the Porter escape, it is situated in between supers with honey to be extracted and those that will be left with the bees. Both models work in the manner detailed above, permitting bees to move down to lower supers and out of the hive, but not back up into the honey supers. Porter escapes cost less than their triangular cousins; however, their design, which enables only one bee at a time to exit, can cause clogging, making evacuation of the super more drawn out.

Triangular bee escape

If any brood is present in the frames, a good number of the bees will not leave. You can prevent the queen from laying brood in honey supers by installing queen excluders in those hives for which you intend to use bee escapes come extraction time. When using bee escapes, make sure all upper entrances to hives, such as those found on the side panels of inner covers, are closed off and that the outer cover is firmly in place. Over the course of several days, opportunistic bees from other colonies can rob supers full of honey while you are waiting for your bee escape to take full effect. When you return to the hive to gather up the honey supers, simply brush off any bees still clinging to the frames. Remove the full super, cover the top and bottom with a bee-tight cover, place it on whatever means of transport you will be employing, and proceed until finished. Remove the bee escape, replace the inner cover to the hive, and top off with the outer cover.

Bee escapes are most effective when evenings are still somewhat cool, which entices bees to move down to the brood box to keep warm. If you find you need to harvest honey while nights are still warm, bee escapes may not be the best means of getting the bees to evacuate. Bees in upper supers have no real incentive to move down then. On those occasions, you want to look into other means of evacuating supers.

Fume Board

A final option for removing bees from supers intended for extraction is the use of a fume board. These boards employ the use of an airborne repellent to move bees off of supers. They resemble outer covers, with the exception that their interior is lined with an absorbent material, such as flannel. To use a fume board, begin by smoking the hive gently. Remove the outer and inner cover. Sprinkle about 1 tablespoon (you needn't measure this, just be judicious and eyeball the amount you dispense) of your repellent of choice (more about repellent options ahead) evenly over the absorbent side of the fume board. Place it atop the topmost honey super, and leave it in position for three to four

Fume board

minutes. This is usually all it takes for the bees to move down into the next super. Take off the fume board, and confirm that the bees have left the super. Remove the super, seal it on the top and bottom with a bee-proof cover, and move it to whatever means of transport you will be using. Move on to the next super, repeating the process until complete. A sunny day is ideal for this operation, as the sun's warmth enhances the repellent's abilities.

There was a time when the only repellent options were highly noxious, hazardous to the bees and environment at large, and, well, just plain smelly. Included in this category are repellents such as benzaldehyde and butric anhydride (in the form of "Bee Go"). Not only do the bees loathe the smell of these chemicals, so will you. If you get a drop on any of your protective gear, consider them history. You'll never want to smell them again. Fortunately, a new repellent utilizes natural oils and herbs with results just as good as its toxic forebears. Known as Bee Quick, this natural repellent safely evacuates bees (who find its scent distasteful, but not harmful), while smelling pleasantly of almonds to human noses.

STYLES OF HONEY

Most commonly found in its runny, liquid incarnation, there are actually several styles of the beloved sweet stuff. For the beekeeper just getting started, you'll probably want extracted honey, as navigating both the equipment and the technique used in harvesting comb honey are concepts better grasped once the more rudimentary aspects of beekeeping are learned. I recommend saving comb honey extraction for your second year.

Comb honey

Comb Honey

This style of honey is extracted exactly as the bees made it, comb and all. To collect the comb, beekeepers may either simply cut the entire portion out of a frame and then section it up (remember to remove the foundation wire), or use special frames intended for the formation of individualized sections of comb and honey. If you are using a top bar hive, comb honey is your only option. Carefully remove the capped honeycomb, cut it free of the top bar and move on to your packaging options.

Chunk Honey

A little bit of both, chunk honey combines comb and extracted honey. A sizeable bit of comb is added to a jar and then topped off with liquid honey. Chunk honey offers something for every honey-loving palate.

Chunk honey

Extracted Honey

Perhaps the best known form of honey, extracted honey is the liquid portion removed from uncapped frames. By far the most economical means of production, extracted honey allows for comb-filled frames to be returned to the hive after extraction, saving the bees the work of fashioning it again.

Extracted honey

Whipped Honey

Honey is a completely saturated solution, meaning that it contains more dissolved sugar than can normally remain in solution form. Accordingly, this solution is quite unstable and, in time, looks to rebalance itself. Over time, the excess glucose in honey separates out, forming granulated crystals. Whipped honey is simply a method of controlling this crystallization, rendering a smooth, spreadable substance out of the granules. If you have honey that has crystallized, becoming a thick, solid, grainy substance, mix one part of it with nine parts liquid honey to create homemade whipped honey.

Whipped honey

Profile of Bee Advocates

John and Karen

As chefs, John and Karen are acutely aware of just how essential the honeybee is to the work that they in turn do. Serving as Executive Chef and Executive Pastry Chef respectively, the husband and wife team are the backbone of Town House, a forward-thinking American cuisine restaurant in tiny Chilhowie, Virginia, specializing in stunningly gorgeous and unexpectedly paired dishes utilizing local and seasonal ingredients.

When Karen became aware of the dangers imposed by Colony Collapse Disorder, she was motivated to spread awareness not just to her restaurant's patrons (via honeybee fundraising dinners), but to other chefs, as well. Considering that honeybees are responsible for pollinating over 90 U.S.-grown crops, chefs could pretty much consider attention to the plight of the honeybee a form of job security. Paying attention, and then acting on it, is exactly what Karen has done. As she describes it, "Respecting food, plants, animals, equipment, space, and colleagues is a huge part in the differentials between being a cook and being a chef."

While Karen and John don't tend hives themselves, they are huge beekeeping advocates, having adopted a hive from The Chef's Garden project, based in Huron, Ohio. A family-run farm, The Chef's Garden knows firsthand the important role bees play in agriculture. As part of their commitment to sustainable growing practices, they have created a project, dubbed "Bee Sustainable," that "provides bees with a consistent source of healthy nectar to combat Colony Collapse Disorder."

Karen and John's personal mission, reflected in the foods they serve at Town House, in addition to the benefit dinners they sponsor and the awareness they are raising amongst chefs, is highly laudable. The couple aspire to "support local farmers, spread awareness of the decline in honeybee populations, advocate on the importance of whole foods and the importance of minimizing refined sugars as part of a healthy diet, raise awareness about the benefits of consuming raw foods, and inspire pertinent questions regarding where one's food comes from." As chefs on a food awareness-raising mission, John and Karen know all too well that "You are what you eat. Respect Mother Nature."

Extraction Action

2 Position a frame over the wax cappings basin.

Most of these basins will have the point of a small nail sticking out from a strip of wood running the length of the tub. Holding a frame from the one end of the top bar, place the opposite end onto this nail point. The nail helps to hold the frame in place as you slice off the wax. It also helps you swing the frame back and forth, accessing both sides with ease. Lacking this, simply position your frame over whatever you are using to cap wax cappings, and hold the frame as steady as possible.

Once you've determined an appropriate extraction location, gathered up all the necessary equipment (and extra sets of hands, if at all possible!), and evacuated the bees from the hive, now it's time to get down to business. The amount of time the entire endeavor will take depends completely on the number of frames you'll be extracting from. Plan on at least several hours from first frame uncapped to clean-up completion.

1 Begin by preheating your uncapping knife.

If you are using an unheated knife, remember to dip it in hot water between uses and wipe off any water remaining on the blade. Position a bottling bucket/dispenser topped with a filter/strainer beneath the spigot of the extractor.

3 Working from top to bottom, remove the white cappings with your uncapping knife.

Use a gentle back-and-forth sawing motion, holding the knife against the wax comb as you run it the entire length of the frame. Repeat on the other side. Hold the frame at such an angle that the wax cappings fall down into the tub below.

4 If, after using the knife, any cells of honey-comb remain uncapped, **use the uncapping comb to break them open.**

5 **Place the uncapped frame vertically into the extractor.**

Repeat with the remaining frames until you've used them all or the extractor is filled. Depending on your model, your extractor might hold only two to four frames, or many, many more. Some models are also reversible, wherein baskets inside the extractor pivot the frames, extracting from both sides; with nonreversible models you'll need to manually lift and flip around the frames in order to remove honey from the opposite side. Finally, your model may be hand-cranked or electric. Based on the specifics of the model you are using, fill your extractor evenly (much like a washing machine, if the frames aren't evenly distributed, the extractor will become unbalanced and wobble), spacing frames across from each other to spread out the weight.

6 **Slowly begin to extract, gradually increasing the speed.**

The centrifugal force will begin literally flinging the honey out of the comb and onto the extractor walls. From there, it will drip down and out of the spigot into the bottling bucket/dispenser positioned below. If you are using a nonreversible model, you will need to flip the frames halfway through the process, extract on the other side, and then repeat again on the other side. If you are using a hand-cranked model, you will need to stop at some point and drain off some of the honey into the bottling bucket and filter below; otherwise, it will pool in the bottom, blocking the extractor's ability to spin with ease. Don't wait too long in a two-frame extractor to reverse the frames. The weight of the inner side can cause to emptying side to crack in half. You may have to do the reversing process several times.

9 Clean your extracting equipment thoroughly with hot water and dish soap.

If you borrowed or rented the equipment, you'll want to be sure to return it just as clean as you received it. Once the frames are emptied of honey, put them back into empty supers and return the supers to the hive. Place them on top of an inner cover that has a hole in the center. Put the outer cover on and leave on for three to four days, then remove the cleaned and dried-out supers. The bees will slurp up any remaining droplets.

7 When the spinning becomes audibly lighter and no more honey seems to be coming out of the frames, stop the extractor.

Remove the frames from the extractor, put them back into an empty super, and place the super outside for the bees to clean.

8 Place a lid over the filtered honey bucket to prevent moisture, dust, or other ambient debris from entering, and leave the newly extracted honey to settle for around 24 hours or so.

This gives time for any air bubbles inside the honey to rise to the top of the bottling bucket, keeping them out of jars once bottled.

10 Bottle and label your honey.

Labels are available through beekeeping equipment suppliers, or you can use those found in craft stores or online, through sites such as Etsy.

A variety of honey containers

THE HONEY LARDER

Honey quality changes over time. The temperature at which it is stored, along with how long it is kept, can help to extend its integrity. Ideally, unprocessed, raw honey should be kept below 57°F (14°C). At higher temperatures, honey stored for any lengthy period of time will begin to granulate. While certainly not harmful, granulation can make honey somewhat less attractive, which is really only an issue if you intend to sell it. Light, either natural or artificial, can also compromise honey taste and texture, so store in a darkened area. A basement would be ideal. Lacking that, store it in the coolest area in your home.

A wealth of containers are available for bottling honey. From simple mason jars to small hexagonally shaped glass vessels to plastic honey bears and beyond, come jarring time you won't suffer from a lack of options. If you intend to offer your honey only to yourself, your friends, and your family, then repurposing any clean, crack-free glass container is fine (plastic will work too, provided it contains absolutely no residual smells or residues, as plastic containers sometimes can; give it a thorough cleaning in a dishwasher first). Be certain to use clean, odor-free lids; this is not the place for pickle jar tops, as the scent will permeate the honey.

If you'd like to sell your honey to the general public, you'll need to use newly purchased, sterilized jars. Beekeeping suppliers carry a range of bottle varieties for purchase. You can also find a treasure trove of options from companies selling glass bottles online. One of my favorites is the Muth jar, named after

Cold Storage

Once you have finished extraction for the year, you'll need to store your honey supers in a cold, safe, secure area, far from the reaches of bacteria, fungi, and wax moths. These moths, which we discussed in detail in chapter 9, can seriously wreak havoc in a hive. How to safely store the supers, then?

If you've used a queen excluder, never allowing the queen to lay brood in your honey supers, the moths don't seem to really be a problem. When the season is over, store the extracted honey supers on a flat surface. This can be done either indoors or outdoors; whichever you choose, the environment needs to be cold and exposed to air. To store, stack the empty supers at 90° angles, one atop the other. If storing outdoors, make sure the supers are in a location with an overhead cover, keeping them protected from rain or snow.

Should mice be of concern in your storage area, once you've found a level surface, stack the hives directly on top of each other and cover with an outer cover. The cover keeps out mice while the cold temperatures keep bacteria, fungi, or wax moth larvae (should they have managed to access your supers) from growing. Storing the supers with the frames in transparent trash bags (also in an unheated area) would also work well, as the wax moths don't like light.

its creator, Charles Muth of Cincinnati, Ohio. Square vessels with cork tops and skep scenes embossed onto their fronts, Muth jars harken back to the 1800s, imparting a bit of nostalgia, whimsy, and beauty to modern honey.

No matter what style of bottle you use, remember to let your honey rest for at least 24 hours after extraction. This step allows air bubbles to rise to the surface, preventing them from flowing into the jars and rising to the top, producing foam. Although not in the least bit harmful, the presence of foam could incur a bit of unfounded worry in your customer. Be sure to place lids onto the honey-filled bottles just as soon as you finish adding the honey to them. You want to limit the honey's exposure to air, which could introduce undesirable moisture.

HONEY MONEY

If you plan on selling your honey, keep in mind that honey is a food. As such, it falls under the same regulations governing other foods sold to the public at large. Standards on what is or isn't required will vary widely by locality, so check with your governing food regulatory agency (such as the Department of Agriculture in the U.S., which has

individual state outposts) to learn what rules govern your area. For the most part, though, labels must contain the following information:

- The word "Honey" must be very visible.
- The weight needs to be indicated.
- The method of processing, or lack thereof, should be detailed (this includes designations such as "pasteurized," "unheated," "untreated," "raw," "natural," or "unfiltered").
- Your name and address need to be listed.

When it comes time to start selling, cast your net wide. From Web-based food and craft sites to natural food stores and farmers' markets, many opportunities exist for finding an audience for your precious honey. In my area, a good number of restaurants feature locally produced honey in their dishes. You could even pair up with cottage businesses making homemade honey-based body care products as a supplier. Small boutiques selling food, home, and gift items would be worth contacting, as well. See if you can sell some at your children's school. The options for selling your honey are limited only by your imagination.

WHAT'S IN A NAME

Organic? Orange Blossom? Lavender? In order to call your honey any of these things, or any other highly specific designation, you'll have to prove it—which isn't always easy to do. If you intend to market your honey as coming from a certain flower, you might need to have a pollen analysis performed. It might be obvious that your honey was sourced from heather if the hive happens to

Properly labeled honey for sale

Lavender honey

be sited within a large tract of land planted with heather. Bear in mind, however, that bees travel quite far in their quest for nectar. If you're uncertain what flowers they paid a visit to, then simply calling it "wildflower honey" may be the best option.

Organic certification is even more difficult than varietal designation. Certified organic honey must come from crops that have not been treated with pesticides. Even if you're growing organically, it's entirely possible that your neighbor isn't. Unless you own all of the area within a bee's foraging radius (about 5 miles [8 km]), there's no way to be sure the bees haven't gathered nectar from treated plants. Regulations overseeing organic certification vary widely from one locality to the next. Check with your governing food regulatory agency to determine what is required in your area.

SWEET REWARDS

While honey presents its own tasty appeal, several other treasures can also be gathered from a thriving hive. Beeswax, pollen, propolis, and royal jelly benefit both honeybees and humans alike. Beeswax can be rendered into a multitude of useful items, while the other hive products offer numerous health benefits.

Beeswax is another valuable hive byproduct.

Waxing Poetic

Beeswax has long been treasured for its versatility. Incorporated into everything from lipstick to bullets, the sturdy substance possesses a seemingly inexhaustible fount of uses. I've listed a number of its more well-known applications here:

- Accordion-making
- Archery wax
- Balms and salves
- Basket-making
- Blacksmithing
- Bullet casting
- Candle-making
- Candy-making
- Copper sink sealing
- Cosmetics
- Crayons
- Dental procedures
- Dental products
- Dreadlocks
- Ear candles
- Embalming
- Encaustic painting
- Foundation (bee frames)
- Furniture polish
- Glass etching
- Goldsmithing
- Lubricant
- Musical instrument mouthpieces
- Mustache wax
- Sealant
- Soap-making
- Waterproofing (leather)
- "Whipping" frayed rope
- Wood-waxing

Wound Care

In addition to being both delicious and nutritious, honey has a number of beneficial medicinal properties. As part of its chemical composition, honey contains hydrogen peroxide, as well as a number of antibacterial phytochemicals. Furthermore, the fact that it is such a supersaturated solution creates a high degree of osmolality, meaning that water cannot freely move into an area where honey is present, thereby preventing the spread of microorganisms potentially present in water. These natural antimicrobial properties make honey an ideal candidate for use in wound care.

A dab of honey, added to a dressing, can then be applied directly over a wound. The amount of honey to be applied depends largely on how intensely the wound is weeping fluid; the greater the weeping, the greater the amount of honey necessary. Honey is also quite beneficial when applied to minor burns. Of course, those infected wounds or those of extreme severity should be examined by a physician.

HEALTH FROM THE HIVE

We all know that honey tastes good—really good. What's even better is how good it is for us. Honey, as well as other products from the beehive, are nutrient powerhouses, chock full of goodness. Here's a look at some of the vitamins, minerals, and other nutritional offerings available from consuming honey, pollen, propolis, and royal jelly.

Honey

Vitamins A, C, D, E, K, beta-carotene, B-complex vitamins (all of them), calcium, chlorine, copper, iodine, iron, magnesium, manganese, phosphorus, potassium, sodium, sulfur, and zinc

Pollen

All essential amino acids, vitamins C and E, beta-carotene, B-complex vitamins (all of them), calcium, copper, iron, magnesium, manganese, phosphorus, potassium, zinc, trace minerals, rutin, lecithin, and lycopene

Propolis

Vitamin A, biotin, calcium, cobalt, copper, iron, magnesium, manganese, niacin, potassium, phosphorus, riboflavin, silica, thiamine, and zinc

Royal Jelly

Vitamins A, C, D, and E, adenine, biotin, cobalamin, folic acid, inositol, calcium, chromium, iron, magnesium, niacin, phosphorus, potassium, pyridoxine, riboflavin, silicon, sulfur, thiamine, and zinc

Profile of a Beekeeper

Eli

A veritable "bee whisperer," Eli's introduction to honeybees couldn't possibly have been more auspicious. A professional gardener tending residential gardens in the city of San Francisco, he had begun noticing fewer and fewer bees appearing in the gardens he manages. Having heard about Colony Collapse Disorder, he was concerned about the bees' dwindling populations. One day, after seeing three bees on a white Icelandic poppy in one of his gardens, he became "so happy I invited them to tell their friends that I had planted a vegetable garden and to feel free to come pollinate it. A week later while I was working from home I heard a really loud buzzing sound in the front. I thought it was someone using a chainsaw in my front garden, so I ran out to see what was happening, took one look, and ran right back inside. My entire garden was *full* of golden, humming honeybees. At first I did not understand what was happening, so I went online and looked for information and figured out it was a honeybee swarm. I guess they accepted my invitation!"

Now an urban beekeeper with the advantage of a large backyard (his fenceless yard connects with the two homes next door), Eli's bees benefit from city forage offered by flowering trees such as eucalyptus, cherry, and plum. His bees also gather nectar and pollen from area yards and parks, bringing food gathered from rosemary, wild blackberry, poppies, salvias, and ceanothus home to the hives. Tending anywhere from two to five hives (depending on the number of swarms he catches in a year), this amateur childhood entomologist has finally fulfilled his lifelong dream of being a "herder of a million stinging insects." Aside from mindfulness surrounding the essential role honeybees play in pollination, Eli keeps honeybees for the sheer joy it offers him, his pleasure in giving honey to friends, the close proximity to nature, and the process of working with the bees, which he finds quite soothing.

Eli's acquired apiary wisdom is worth passing on. Here are his four biggest recommendations for successful beekeeping:

"**Meet other beekeepers.** They are an excellent resource. We love bees, love to talk about bees, love to look at bees, love to have any excuse to be around bees, so we are very likely going to want to help out with questions.

Let the bees do their thing. They know what they are doing. Just try to give them enough space in the hive for the queen to lay eggs and the workers to store nectar so they don't feel crowded and decide to leave.

Don't buy used equipment, even if it is a great deal. I introduced foulbrood and mites into my hives this way and forever regret it.

I take a lot of photos while I work. I find them very helpful after I close up the hive. I see things I did not notice while working the hive. It teaches me to be observant."

Chapter 11
Recipes

Honey lovingly extracted from your own hives is quite literally one of life's sweetest pleasures. While it's certainly tempting to just eat it straight out of the jar, an infinite world of possibilities for transforming it into sweet and savory delicacies awaits the home cook. I've shared some of my favorite recipes here, covering the culinary repertoire "soup to nuts." You'll even find two beverage ideas, including a fragrant nighttime tipple. I trust they'll soon become some of your most favorite go-to honey-starring dishes.

Adding fresh herbs and spices to honey is a quick and easy way to introduce a world of flavor variations to the ambrosial nectar. You can vastly increase your culinary repertoire with infused honeys in your pantry. They're also simple, yet fantastic, ideas for gifts that your friends and family will no doubt treasure deeply. *Yield: 1 cup (8 ounces)*

YOU WILL NEED:

1	cup honey
	Herb or spice infusing agent of choice*

Herbal infusing options (fresh herbs are preferable): basil, chamomile, lavender, lemon balm, marjoram, peppermint, rosemary, rose petals, sage, spearmint, tarragon, and thyme

Spice infusing options (whole, not ground, spices are essential): allspice berries, anise seeds, cardamom pods, cinnamon sticks, citrus peel, whole cloves, crystallized ginger, fresh gingerroot slices, star anise, and vanilla bean pod

Infused Honey

TO PREPARE:

1. Sterilize either two 4-ounce or one 8-ounce jar(s) by submerging into boiling water for 30 to 40 seconds. Using a cloth, dry completely, leaving no traces of water whatsoever.

2. Place the infusing agent of choice into the jar(s). If using fresh herbs, use 1 or 2 sprigs, making sure they are completely dry (any beads of water can cause the honey to mold and spoil); if using spices, use 1 to 2 teaspoons, depending on the intensity of flavor desired.

3. Warm the honey in a stainless-steel pot over medium-low heat until it moves easily in the pot, appearing completely "runny." Don't allow the honey to boil, only to be fully warmed.

4. With the aid of a funnel, pour honey into each jar, completely covering the infusing agent.

5. Allow jar(s) to cool at room temperature. Once fully cooled, place a lid or cork (depending on bottle being used) over the jar's opening.

6. Store in a cool, dark area, such as a pantry or basement. Allow to infuse for at least one week, preferably two, before use. Use within one year.

7. If you prefer a clear presentation, you may strain off the jar's contents after the infusing period, composting the solids and rebottling the infused honey.

Holiday Rounds

Whipped up one weekend afternoon for a honeybee benefit potluck, these holiday rounds proved to be an absolute hit. One friend, a professional baker, declared them to be "the best thing I've eaten in a long time." Fun and festive, these treats work equally well for a posh soiree or a teatime snack. *Yield: 3 to 4 dozen*

YOU WILL NEED:

1 cup walnuts

1 cup pecans

½ cup almonds

1 cup dried apricots, chopped

¼ cup butter (½ stick)

¼ cup honey

½ cup dried cranberries

2 tablespoons fresh rosemary, finely chopped

½ teaspoon nutmeg (freshly grated, if possible)

1 teaspoon sea salt

Black pepper, to taste

TO PREPARE:

1. Pulse the walnuts, pecans, almonds, and apricots in blender until the pieces form crumbles (don't overdo it, or you'll end up with a nut and fruit butter).

2. Melt the butter in a stainless-steel saucepan over medium heat. Add the nut mixture. Stir for about a minute, then add the honey. Cook for three to four minutes, stirring frequently.

3. Add the cranberries, rosemary, nutmeg, salt, and pepper. Cook two to three minutes, stirring frequently, until mixture is sticky.

4. Remove from the heat and let cool for five to 10 minutes.

5. Form the mixture into small balls, just under 1 inch (2.5 cm) in diameter. Enjoy at room temperature or store in an airtight container in the refrigerator and consume within two weeks.

I long ago stopped purchasing store-bought salad dressings. Using fresh quality ingredients, I found I could quickly put together my own dressings, spending considerably less and getting far more in terms of flavor. This vinaigrette uses grain mustard but would work equally well with Dijon. Toss it with some arugula, use it as a flavorful marinade for chicken, or pair it with cold pasta and chopped tomatoes for an easy summer salad. *Yield: 2 cups (16 ounces)*

YOU WILL NEED:

1	cup olive oil
¼	cup honey
3	tablespoons grain mustard
3	tablespoons apple cider vinegar
¼	teaspoon garlic powder
	Sea salt
	Black pepper

TO PREPARE:

1. Place the olive oil, honey, mustard, vinegar, and garlic powder into a lidded container, such as a glass mason jar. Top with lid and shake until well combined. Add salt and pepper to taste and shake again.

2. Use immediately or store covered in the refrigerator. Olive oil will become solid upon refrigeration, so either bring to room temperature before use by leaving it on the kitchen counter or microwaving briefly. Shake to recombine.

Grain Mustard Honey Vinaigrette

This is the perfect dish to make when root vegetables are available in abundance. Honey works so well here, drawing out the natural sweetness of the vegetables, playing expertly with the fragrant aromatics of rosemary and sage. Serve this dish alongside roasted chicken or turkey, or with wild rice pilaf and sautéed greens for a vegetarian meal.

Yield: Six to eight 1/2-cup servings

Roasted Root Vegetables with Honey & Herbs

YOU WILL NEED:

1 large turnip, cubed

1 large beet, cubed

2 large Yukon gold or other medium-starch potatoes, cubed

1 pound carrots, peeled and cubed

1/2 cup honey

1/3 cup extra-virgin olive oil

1/3 cup rice wine vinegar

Sea salt

Black pepper

2 tablespoons fresh sage, finely chopped

2 tablespoons fresh rosemary, finely chopped

TO PREPARE:

1. Preheat the oven to 400°F (204°C).

2. In a large mixing bowl, toss the cubed turnips, beets, potatoes, and carrots with the honey, olive oil, rice wine vinegar, sea salt, and black pepper.

3. Spread the mixture evenly across a 12 x 17-inch (30.5 x 43.2 cm) rimmed, oiled baking sheet. Bake for 30 minutes.

4. Remove the baking sheet from the oven. Using a spatula, turn the vegetable cubes over. Spread the mixture evenly again, return the sheet to the oven, and bake an additional 20 minutes.

5. Remove the baking sheet from the oven. Using a spatula, toss in the sage and rosemary. Return the baking sheet to the oven, and bake an additional 10 minutes.

6. Remove the baking sheet from the oven. Allow to cool several minutes. Transfer the roasted vegetables to a serving dish, and serve immediately.

Chestnut Soup with Honey

Silky, delicate, and layered in levels of flavor, this soup will leave you with a serious urge to lick the bowl clean. Chestnuts, with their sweet and subtle flesh, are ideal when what you want is a creamy soup. Paired here with honey, potatoes, and the unexpected spark of fennel seed and ginger, this dish warms the belly and delights the senses. A great choice to serve as a first course. *Yield: Six 1-cup servings*

<table>
<tr><td>YOU WILL NEED:</td><td>1</td><td>pound fresh chestnuts (or 2 cups roasted, peeled, jarred chestnuts)</td></tr>
<tr><td></td><td>2</td><td>Yukon gold potatoes</td></tr>
<tr><td></td><td>2</td><td>tablespoons extra-virgin olive oil</td></tr>
<tr><td></td><td>2</td><td>tablespoons butter</td></tr>
<tr><td></td><td>½</td><td>onion, diced</td></tr>
<tr><td></td><td>1</td><td>tablespoon ginger, peeled and finely minced</td></tr>
<tr><td></td><td>1</td><td>tablespoon fennel seed</td></tr>
<tr><td></td><td>4</td><td>cups vegetable or chicken stock</td></tr>
<tr><td></td><td>2</td><td>cups milk</td></tr>
<tr><td></td><td>⅓</td><td>cup honey</td></tr>
<tr><td></td><td>3</td><td>tablespoons lemon juice</td></tr>
<tr><td></td><td>¼</td><td>teaspoon ground nutmeg</td></tr>
<tr><td></td><td></td><td>Sea salt</td></tr>
<tr><td></td><td></td><td>Black pepper</td></tr>
<tr><td></td><td>2 to 3</td><td>tablespoons brandy (optional)</td></tr>
<tr><td></td><td>2</td><td>tablespoons chopped green onion (optional)</td></tr>
</table>

TO PREPARE:

1. Preheat the oven to 420°F (216°C). Fill a medium bowl with water, and add a pinch of salt. Place each chestnut flat side down on your cutting board and, using a serrated knife, score with an X. Place the chestnuts in the water, and set aside. Omit this step if using jarred, peeled chestnuts.

2. Peel and quarter the potatoes. Toss the potatoes with the olive oil and place them on a small baking sheet. Put in the oven and bake for 30 minutes.

3. At the end of 30 minutes, remove the chestnuts from the water, put them flat side down on a small baking sheet, and place the sheet in the oven alongside the potatoes. Cook the potatoes and chestnuts for 20 more minutes, then remove both pans from the oven. Set the potatoes aside.

4. Place the chestnuts in a dry dish towel, wrap it around them, and then squeeze the towel to crack the shells. Set aside to cool for five minutes.

5. When the chestnuts have cooled enough to handle, peel off the shells and set aside the nut-meat in a separate bowl.

6. Melt the butter in a large stainless-steel saucepan over medium heat. Add the diced onions. Sweat the onions for several minutes, until translucent.

7. Add the ginger, and stir until fragrant. Add the fennel seeds, stirring until fragrant. Add the stock and the milk. Reduce the heat and bring to a gentle simmer. Add the chestnuts, potatoes, honey, lemon juice, nutmeg, sea salt and black pepper to taste, and brandy if desired. Simmer for five minutes. Remove from heat and let cool five to seven minutes.

8. Working in batches, place the soup in a blender, and puree until smooth and creamy.

9. Ladle the soup into bowls, and serve. Top with green onion, if desired.

Honeyed Prawns & Polenta

This recipe presents a fun variation on the traditional dish of shrimp and grits associated with the southeastern coastal regions of the United States. Polenta, thyme, capers, feta cheese, and currants update the classic with Mediterranean flavors. The inclusion of honey injects just the right amount of sweetness to complement the dish's many savory flavors. I encourage the use of sustainably raised and harvested prawns here, as they exact a considerably smaller toll on our global aquatic ecosystems.

Yield: Four ³/₄-cup servings

YOU WILL NEED:

For the prawn marinade:

4	tablespoons tomato paste
3	tablespoons honey
1	tablespoon lemon juice
1	tablespoon extra-virgin olive oil
1	teaspoon hot sauce
¹/₂	teaspoon garlic powder
	Sea salt
1	pound large, peeled, and deveined prawns

For the polenta:

1	cup polenta
1¹/₄	cups (10 ounces) corn kernels, fresh or frozen
¹/₄	cup currants
1	tablespoon honey
1	tablespoon butter
1	teaspoon fresh thyme, or ¹/₂ teaspoon dried
	Sea salt
	Black pepper

For the topping:

¹/₃	cup feta cheese, crumbled
2 to 3	tablespoons fresh parsley, finely chopped
2	tablespoons capers
	Hot sauce (optional)

1. Preheat the oven to 400°F (204°C). Butter an 8 x 10-inch (20.3 x 25.4 cm) baking dish. Oil a rimmed baking sheet with olive oil. Set the pans aside.

2. To make the marinade: Combine the tomato paste, honey, lemon juice, olive oil, hot sauce, garlic powder, and a pinch or two of salt in a medium bowl. Toss the prawns with the marinade. Set the bowl aside, giving it a stir every few minutes.

3. For the polenta: Bring 3 cups cold water to a boil in a medium stainless-steel pot. Add the polenta. Cook seven minutes, stirring often. Remove the pot from the heat, cover, and let sit for 10 minutes.

4. In a large bowl, mix the corn (completely thawed if frozen), currants, honey, butter, and thyme. Stir in the polenta, and season with salt and pepper to taste.

5. Add polenta mixture to the buttered baking dish. Spread evenly with a spatula, and place in the oven to bake for 35 minutes.

6. Spread the prawn mixture evenly onto the oiled baking sheet. After the polenta has baked for 25 minutes, add the prawns to the oven, alongside the polenta. Place both pans on same rack if possible; otherwise, place the baking sheet on the lower rack.

7. Bake the prawns for five minutes. Remove the baking sheet from the oven, and flip all of the prawns over. Return to the oven and bake an additional five minutes, until the polenta is slightly browned and the prawns are pink and firm.

8. Remove the prawns and the polenta from the oven. To serve, scoop individual servings of the polenta onto plates. Lay a few prawns over each serving, then top with crumbled feta, fresh parsley, and capers. Drizzle with a few lashings of hot sauce, if desired.

Honey Ice Cream

Honey and ice cream were pretty much made for each other. Floral and fragrant, the honey kisses the custardy mixture with just the right amount of sweetness. A treasured warm-weather treat throughout Europe, honey ice cream showcases the best of summer with every frozen spoonful. *Yield: 5 cups*

YOU WILL NEED:

2 cups milk
$^2/_3$ cup honey
$^1/_2$ teaspoon sea salt
2 large eggs
2 cups heavy cream
1 tablespoon vanilla

TO PREPARE:

1. The day before you would like to begin, place the bowl of an ice cream maker in the freezer to chill.

2. Heat the milk gently in a medium stainless-steel pot over medium-low heat, about four to five minutes. Gradually whisk in the honey and salt.

3. Beat the eggs in a small bowl. Remove about ½ cup of the milk mixture from the pot and whisk it slowly into the eggs. Once fully incorporated, stir the egg mixture into the remaining milk mixture in the pot. Heat the milk over medium-low heat for an additional four to five minutes, stirring constantly to prevent scorching.

4. Remove the pot from the heat, and transfer its contents to a glass, metal, or ceramic bowl. Allow to cool completely.

5. Once the milk mixture has cooled, whisk in the heavy cream and vanilla. Place the bowl in the refrigerator, and chill until cold throughout.

6. Transfer the mixture to the bowl of an ice cream maker, and process according to the manufacturer's instructions.

During the warmer months, chances are, if I'm not eating eggs for breakfast, I'm feasting on this granola. Crunchy, spicy, and sweet, this wholesome blend makes for a satisfying, and filling, cereal. Of course, it's just as delicious as a midday treat or a late-night snack! *Yield: 8 cups*

YOU WILL NEED:

- 3 cups old-fashioned rolled oats
- 1 cup raw almonds
- 1/2 cup roasted pumpkin seeds, unsalted
- 1/2 cup roasted sunflower seeds, unsalted
- 3/4 cup unsweetened dried coconut flakes
- 2/3 cup honey
- 1/2 cup extra-virgin olive oil
- 1/4 cup light brown sugar
- 2 teaspoons ground cinnamon
- 2 teaspoons ground cardamom
- 1 teaspoon sea salt
- 1/2 cup raisins
- 1/2 cup dried apricots, chopped

TO PREPARE:

1. Preheat the oven to 300°F (149°F). Oil a large, rimmed baking sheet with olive oil. Set aside.

2. Combine the oats, almonds, pumpkin seeds, sunflower seeds, coconut, honey, olive oil, brown sugar, cinnamon, cardamom, and salt in a large bowl. Stir to coat completely. Spread the mixture evenly onto the prepared baking sheet.

3. Bake for 40 minutes, stirring every 10 minutes.

4. Remove the baking sheet from the oven, cool 10 minutes, and then transfer its contents to a large mixing bowl. Stir in the raisins and apricots.

5. Store in an airtight container, and use within three to four weeks.

Fruit, Nut & Honey Granola

I first encountered an apple cider-based version of this old winter nightcap standby at a local pub. With a bit of tinkering, I came up with the recipe below. The sweetness of the apples, coupled with the honey, offers up a perfect foil to the bite of the whiskey and the tartness of the lemon juice. Whip some up, gather around a woodstove or outdoor fire, and feel the chill of winter slowly melt away. **Yield: One 8-ounce serving**

Hot Cider & Honey Toddy

YOU WILL NEED:

³/₄ cup fresh apple cider

1 tablespoon honey

2 ounces whiskey

2 teaspoons lemon juice

Ground cinnamon

TO PREPARE:

1. Warm the apple cider.

2. Drizzle the honey into the bottom of a cocktail shaker. Add the warm cider, whiskey, lemon juice, and a pinch of cinnamon.

3. Shake vigorously for about 30 seconds. The top of the cocktail shaker may pop off a bit at the beginning as the contents expand from the warmed cider, so you may want to shake over a sink.

4. Strain contents into a glass, and serve.

Honey & Ginger Cold-Fighting Tea

This zesty tea is my absolute go-to elixir of choice should anyone in the house find themselves with a scratchy throat or a burgeoning cough. The heat of ginger and cayenne work to loosen phlegm while honey soothes the irritated lining of the throat. It's also exceptionally delicious and can easily be rendered into an iced version for a refreshing summer pick-me-up.

Yield: Four 8-ounce servings

YOU WILL NEED:	
4	cups water
2	tablespoons fresh ginger, peeled and chopped
2	tablespoons lemon juice
2 to 4	tablespoons honey, to taste
	Cayenne pepper
1 to 2	cloves garlic, minced (optional)

TO PREPARE:

1. Bring the water to a boil. In a ceramic teapot, combine the ginger, lemon juice, honey, and a pinch of cayenne. For a dose of added medicinal value, add fresh garlic, if you like.

2. Pour the boiling water over the ginger mixture. Stir, and allow the mixture to steep for about 10 minutes. Strain off the liquid through a fine mesh sieve, pour tea into mugs, and enjoy.

Notes

BEE CARE CHECKLIST

Care for honeybees is largely influenced by the changing seasons. Knowing what to do and when is essential for proper stewardship of your buzzing beauties. Here I've provided a seasonal listing of tasks, duties, activities, and such. I'd suggest making a copy, placing it in a plastic sleeve, and attaching that sleeve to the underside of your hive's outer cover for safekeeping and easy reading.

Spring

EARLY

- Look inside only on warm, sunny days.
- Check for eggs, uncapped and capped brood, and the presence of a queen.
- Begin supplemental feeding if food stores are low; continue until an active nectar flow occurs.
- Remove mouse guard and entrance reducers.
- Install queen excluder, if using, over brood box.
- Add supers as soon as brood box is full of bees.
- Continue adding supers as each new one fills with honey or bees.
- Begin weekly inspections in mid-spring, checking for swarm cells.
- Remove and replace old frames and foundation.
- Make splits as needed.
- Requeen or allow natural supersedure as necessary.

LATE

- Reverse hive bodies.
- Determine what swarm-capturing method you will use, in the event that it should occur.
- If an intense nectar flow happens, a late spring honey extraction may be necessary.
- Plant honeybee-loving plants near your hive and around your property.
- Examine varroa mite population and treat as desired.

Summer

- Inspect the hive every week, monitoring the queen's activities, honey production, and presence of any swarm cells. Look for eggs, larvae, and other indications of active laying.
- Continue adding supers as needed, both for honey as well as for brood.
- Be prepared to extract, most likely in late summer.
- Be prepared to offer sugar syrup should your area experience atypical weather, such as a drought, unseasonably cool temperatures, or excessive rainfall, as such situations may affect the bees' access to nectar sources.
- Install the entrance reducer if robbing evidences or wasps or yellow jackets attempt to gain entry.
- Examine varroa mite population and treat as desired.

Autumn

- Check in on the hive every two weeks. Stop inspections completely once the temperature drops below 50 to 55°F (10 to 13°C) during the day.
- Check for the presence of a queen. Look for eggs, capped and uncapped brood, and a good laying pattern.
- Check on capped honey stores. If they are low, you will need to feed supplementally. Stop all feeding by late autumn.
- Tilt hives slightly forward to allow rain and snow to run off.
- Ventilate inner cover.
- Remove queen excluder, if using.
- Remove hive-top feeder, if using, and replace inner cover.
- Install a mouse guard.
- Install entrance reducer, if not already using.
- Secure hives with either a large, heavy rock atop the outer cover or a strap running the circumference of the hive (from side to side, not from entrance to rear).
- Wrap hives with black roofing tar paper if you live in an area experiencing severe winters, where temperatures remain below freezing for months.
- Store empty honey supers in a secure location.

Winter

- Check the bee yard regularly for damage caused by storms or predator animals.
- Check that mouse guards are still in place.
- Check food stores by lifting hive to determine weight. Consider supplemental feeding if necessary.
- If no water source is available nearby (such as a creek or river), provide unfrozen water; be sure to place it in a sunny location.
- Gently tap on the side of the hive to hear buzzing (an indication that the hive is alive).
- Order more frames and supers, or other equipment, as needed.
- Continue reading and learning about honeybees.

Resources

SUPPLIERS

Bee Commerce
11 Lilac Lane
Weston, CT 06883
www.bee-commerce.com

Betterbee
8 Meader Road
Greenwich, NY 12834
(800) 632-3379
www.betterbee.com

Brushy Mountain Bee Farm
610 Bethany Church Road
Moravian Falls, NC 28654
(800) 233-7929
www.brushymountainbeefarm.com

Dadant and Sons
51 South 2nd
Hamilton, IL 62341
(888) 922-1293
www.dadant.com

Long Lane Honey Bee Farms
14556 N. 1020 E. Road
Fairmount, IL 61841
www.honeybeesonline.com

Mann Lake
510 S. 1st Street
Hackensack, MN 56452-2589
(800) 880-7694
www.mannlakeltd.com

Miller Bee Supply
496 Yellow Banks Road
North Wilkesboro, NC 28659
(888) 848-5184
www.millerbeesupply.com

Rossman Apiaries
P.O. Box 909
Moultrie, GA 31776-0909
(800) 333-7677
www.gabees.com

The Walker T. Kelley Company
P.O. Box 240
807 W. Main Street
Clarkson, KY 42726
(800) 233-2899
www.kelleybees.com

Western Bee Supplies
P.O. Box 190
5 9th Avenue E.
Polson, MT 59860
(800) 548-8440
www.westernbee.com

National Bee Supplies
Merrivale Road
Exeter Road Industrial Estate
Okehampton, Devon EX20 1UD
United Kingdom
+44 (0) 1837 54084
www.beekeeping.co.uk

E. H. Thorne Beehive Works
Wragby
Market Rasen LN8 5LA
United Kingdom
+44 (0)1 673-858-555
www.thorne.co.uk

HoneyBee Australis
P.O. Box 298
Ipswich, Queensland 4305
Australia
(07) 3495 7095
www.honeybee.com.au

Pender Beekeeping Supplies
28 Munibung Road
Cardiff, NSW 2285
Australia
(02) 4956 6166
www.penders.net.au

PERIODICALS

American Bee Journal
www.americanbeejournal.com

Bee Craft
www.bee-craft.com

Bee Culture
www.beeculture.com

Beekeepers Quarterly (UK)
www.beedata.com/bbq.htm

ORGANIZATIONS

**American Association of
Professional Apiculturists**
www.masterbeekeeper.org/aapa/

American Beekeeping Federation
www.ABFnet.org

**International Federation of
Beekeepers' Associations**
www.apimondia.org

**African Beekeeping Resource Centre
(Kenya)**
www.apiconsult.com

**Canadian Association of Professional
Apiculturists**
www.capabees.com

British Beekeepers' Association
www.britishbee.org.uk

**International Bee Research
Association (UK)**
www.ibra.org.uk

**National Beekeepers' Association
of New Zealand**
www.nba.org.nz

WEBSITES

Bee Source
www.beesource.com
Forums and international
supplier listings

The Honeybee Project
www.thehoneybeeproject.com
A crossroads of education, science,
the arts, and technology all in
service to children and the honeybee

Honeybee News
www.honeybee-news.com
International news, interviews,
and helpful links

**Harry H. Laidlaw Jr. Honey
Bee Research Facility**
www.beebiology.ucdavis.edu
Research facility at the University
of California at Davis

**Texas A&M University
Honey Bee Information**
www.honeybee.tamu.edu
Links to state beekeepers'
associations

Glossary

Africanized Honey Bee. A race of honey-bee, originating in Africa, that crossbred with feral bee populations in Brazil during the 1950s and has since crossed into North America. These bees are very defensive and known for their aggressive behavior.

Apiary. Known also as a "bee yard," the apiary is the specific physical site of one or more beehives.

Bee brush. A long-handled tool with gentle bristles used for delicately removing bees from a surface

Bee escape. A device inserted between honey supers and brood chambers to remove bees from those supers bound for honey extraction

Bee space. The area around a hive that bees naturally maintain and move between; used in determining where comb will be built. Bee space measures at least ³/₈ inch.

Beeswax. The product excreted by honeybees via glands in their abdomen. Used in building comb that will house brood and food.

Bee yard. Also referred to as an "apiary," the bee yard is the specific physical site of one or more beehives.

Bottom board. The lowest part of a hive, essentially its "floor"

Brood. The term used to collectively describe all phases of immature, developing bees: eggs, larvae, and pupae

Brood box. Also referred to as a "brood chamber," "hive body," or simply a "deep," the brood box is where the queen lives, developing bees are raised, and a large part of the hive's activity takes place.

Burr comb. Known also as "brace" comb, this type of comb is built between frames or between frames the hive's woodenware. Burr comb must be removed by the beekeeper before frames may be removed and manipulated.

Cleansing flight. A flight taken during warm breaks in otherwise cold weather conditions wherein honeybees relieve themselves of accumulated bodily excrement after an extended period of confinement

Cluster. A round, football-shaped formation that honeybees fashion around the queen during cold-weather months. Used to generate and maintain heat inside the hive.

Drawn comb. Comb with cells built by honeybees onto sheets of wax foundation

Drone. A male honeybee

Egg. The first stage in a bee's three-phase development

Entrance reducer. A device inserted into the hive's entrance used to prevent cold weather, robbing bees, or pests from gaining entry into the hive, thereby making the hive easier to defend

Extractor. A device, either manual or electric, used to remove liquid honey from wax cells

Feeder. Any of a number of devices used for providing sugar syrup to bees

Fondant. A solid emergency food made of sugar, water, corn syrup, and a thickening agent

Foulbrood (AFB & EFB). Bacterial diseases affecting brood. Can be life-threatening to entire hive if left unchecked. AFB is highly contagious, and any affected equipment must be burned to contain spread of the disease.

Foundation. A sheet, made of either beeswax or plastic, embossed with hexagonal cells that is positioned between the sides of a frame. Bees will build wax cells onto the hexagonal imprints, developing "drawn comb."

Fume board. A device used to quickly remove bees from supers bound for honey extraction

Frame. A rectangular structure made of either wood or plastic, which hangs from the interior of a super, upon which comb is built. An integral component of a Langstroth hive.

Grease patty. A mix of vegetable shortening and granulated sugar, often including essential oils, fed to honeybees to control tracheal mite populations

Hive body. Also referred to as a "brood box," "brood chamber," or simply a "deep," the hive body is where the queen lives, developing bees are raised, and a large part of the hive's activity takes place.

Hive tool. A device that is flat on one end and curved on the opposite, used in opening and manipulating supers and frames

Honey. The by-product resulting from the dehydration and enzymatic alteration by honeybees of plant nectar. A food source for both honeybees and humans.

Honey flow. The seasonal and regional-specific periods of time when flowering plants are producing abundant amounts of nectar. This period influences the behavior of both bees and beekeeper alike.

Inner cover. A wooden component of the hive, separating the outer cover, or uppermost portion, of the hive from the supers below

Integrated pest management (IPM). A style of pest control encouraging the use of non-chemical practices and preventative care to optimize hive health

Langstroth hive. The most commonly used model of beehive. Composed of removable frames and wooden boxes.

Larva. The second stage in a bee's development. Larvae are white, shiny, grub-like organisms.

Mouse guard. A metal device with perforated openings installed over a hive's entrance during cooler months to keep mice out while allowing bees to come and go

Nectar. The sweet, sugar-rich liquid secreted by certain plants to attract insects for pollination; gathered by honeybees for transforming into honey

Nosema. A digestive disorder of adult honeybees caused by the protozoan *Nosema apis*

Nuc. An abbreviation for "nucleus," a small hive consisting of a queen, brood, and worker bees used for beginning new colonies

Package bees. A mesh-screen shipping container of several pounds of bees, including a mated queen and a feeder can. Used for creating new colonies.

Pheromone. An air- or contact-borne hormonal chemical excreted by bees to arouse certain behaviors in other bees. The "alarm" pheromone, indicating danger, "queen substance" pheromone, indicating the queen's presence in the hive, and "Nasonov gland" pheromone, indicating where a hive is located are three examples of the numerous pheromones used by honeybees to exchange information and influence behavior.

Pollen. The dusty, powdery substance produced by the male reproductive cells of flowers and used by honeybees as a source of protein

Pollination. The act of moving pollen from a plant's anthers (pollen-producing part of a plant) to its stigma (pollen-receiving part of a plant). Honeybees, due to their physiology, are remarkably adept at pollination.

Propolis. The sticky, resinous fluid excreted by plants and gathered by honeybees; used to seal up any cracks, crevices, or openings within the hive to keep drafts and pathogens out of the hive

Pupa. The third and final stage of development in a honeybee's metamorphosis from egg to adult bee

Outer cover. The uppermost portion of a hive. Often referred to as a "telescoping" outer cover, the edges of this covering extend beyond the supers beneath it, permitting rain and snow to fall out and away from the hive.

Queen. A sexually developed female bee that lays eggs. Her presence, or lack thereof, regulates activity within the hive.

Queen cage. A small, screened box used to house a queen while in transit, as well as to introduce her gradually to a colony

Queen cell. An elongated, peanut-shaped cell, often found on the side of a frame, housing a future queen

Queen excluder. A mesh screen used to prohibit upward movement by a queen into honey supers while permitting access to worker bees. Used to keep brood and honey separate.

Requeen. The process of artificially introducing a new queen to a colony, wherein the former queen is replaced

Robbing. The act of one hive stealing food from a neighboring hive. Other insects such as wasps can also perform robbing.

Royal jelly. A highly nutritious substance secreted by worker bees and fed to future queens.

Screened bottom board. A mesh screen used in lieu of a solid bottom board for management of varroa mite populations within a hive

Small hive beetle. A scavenging, small beetle that feeds on brood and food within a hive. Can be of grave concern if not managed properly.

Smoker. A device used during beekeeping inspections and frame removal to transfer cool smoke into the hive, thereby masking the alarm pheromone otherwise given off and calming the bees

Sugar syrup. A fully saturated sugar and hot water mixture fed to honeybees during periods of nectar dearth or initial colony buildup

Super. The general term describing the wooden boxes used to hold frames and house colonies of bees. The combination of various supers forms a hive. Available in deep ($9\frac{1}{2}$-inch), medium ($6\frac{5}{8}$-inch), and shallow ($5\frac{3}{8}$-inch) sizes.

Supersedure. The natural formation of a new queen by worker bees; performed in order to replace an existing queen, usually due to illness, old age, or death

Swarm. A mass of unhived bees. A genetic trait of honeybees, performed for purposes of reproduction and species expansion. Swarming is an activity many beekeepers work fastidiously to prevent, so as to curtail honey production losses.

Tracheal mite. A microscopic parasitic mite that lives in the breathing tubes (trachea) of adult bees

Uncapping fork. A pronged tool used in honey extraction to puncture wax cappings

Uncapping knife. A heated, sharpened tool used to slice off sealed wax cappings from frames in order to extract the liquid honey contained inside

Varroa mite. A small parasitic mite that feeds on the haemolymph (bee blood) of both brood cells and adult bees. Potentially devastating to a hive if populations are not managed.

Veil. A protective netting covering placed over a beekeeper's head and worn during hive inspections. Permits good ventilation and vision while keeping the face safe from potential stings.

Worker. A sexually immature (unmated) female bee, comprising the bulk of a hive's occupants. Worker bees move through a series of duties during their lifetimes, performing all of the hive's nonreproductive functions.

Acknowledgments

An overflowing garden of gratitude is in order for the many talented individuals who pulled together this project.

To the wonderful profilees: Majora Carter, Jon Christie, Chase Emmons, Jenny Kallista, Corky Luster, Megan Paska, Debra Roberts, John Shields, Karen Urie Shields, and Eli Wadley. Thank you all for bringing light to the many different ways we can all enjoy the benefits of honeybees in our lives. Amongst these profilees, I would like to extend particular thanks to Jenny Kallista for her never-ending guidance and boundless enthusiasm as an apiary mentor. Jon Christie and his business, Wild Mountain Apiaries, generously loaned many of the beekeeping items photographed throughout the book.

Cindy Jordan and Laura Blackley were kind to let us photograph both their bee yard and their glorious honey extraction process. Sam Champion of Anarchy Apiaries happily showed off his top bar hive. Great thanks to Mike Tuggle for offering an extra set of seasoned beekeeping eyes and wisdom to the pages of this book.

I applaud Rebecca Springer for waving her copyediting wand over my words with grace and good judgment.

Photographer Lynne Harty never ceased to surprise with her ability to get just the right shot over and over again. Designer Eric Stevens should be lauded for his beautiful design, illustrations, and flawless ability to pull all of the puzzle pieces together in just the right way. Melanie Powell's illustrations put the final important touches of clarity to the nitty-gritty details. Abundant gratitude is offered to Chris Bryant for his invaluable design guidance and styling prowess. Thanks to both Chris and his partner Skip Wade for generously offering up their beautiful home yet again for the demanding but tasty photo shoot.

Heartfelt thanks to Nicole McConville, for believing I had both the skills and gumption to take on this book and the series. I am immeasurably grateful to have such a great friend and editor wrapped up in one convenient package.

To my loving and infinitely patient husband, Glenn, who served as cheerleader, therapist, and official taster through all of this, I appreciate you more than you could ever know.

Finally, special thanks to Meaghan Finnerty, Paige Gilchrist, and Marcus Leaver, for all of the excitement surrounding the series that you have both nurtured and enabled.

Also Available in the Homemade Living Series:

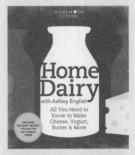

Photo Credits

The pages of this book are richer thanks to the contributed photos. Much gratitude is owed to the following individuals: Sandra Allison (page 83), Leslie Brewer (page 105), Chris Bryant (page 80, 95), Majora Carter (page 64), James Chase (page 65), James Cobb (page 74, 82), Jen Doumen (page 81), Deborah Felkel (page 43), Sara B. Hodge (page 63, 67), Jenny Kallista (page 23), Andrew Langley (page 32), Corky Luster (page 38), Robert Masse (page 58), Enzie Shahmiri (page 13), Krista Theiss (page 32, 56), Mike Tuggle (page 39), Bob Voors (page 32), Eli Wadley (page 113), Summer Walker (page 29), Kenneth Walny (page 62).

Index

Africanized honeybees, 31
Alarm pheromone, 22, 67
allergy, bee-sting, 29
American foulbrood (AFB), 44–45, 85
anatomy, 14–16
antibiotics, 85
ants, 91
autumn, beekeeping tasks during, 81
    checklist, 81, 128
bear fencing, 64, 92
bears, 92–93
bee bread, 46
bee brush, 54
    technique for using, 101
bee escapes, 102–3
bee garden, 95
beekeeping: annual schedule, 74–82
    cost of, 25–26
    regulation of, 27
    time required, 26
beekeeping community, 9
beekeeping equipment, 26, 50–55
    secondhand, 26
    storage of, 55
    supplies, 26
Bee Quick, 103
bee space, 13, 33
bee-sitter, 26
beekeeping history, 13
beekeeping organizations, 25–26, 129
beekeeping suit. See protective clothing
bees. See honeybees
beeswax, 44, 111
beetle traps, 90
Benton cage. See queen cage
blowing, as evacuation technique, 102
Boardman entrance feeder, 48
bottling bucket, 100, 107
bottom board, 34–35
brood, 21, 37
    capped, 71
brood chamber. See hive body
burr comb, 53, 69, 72
capping pattern, 71

castes, 18, 21
Chalkbrood, 86
checklist, bee care, 128
Chestnut Soup with Honey (recipe), 120–21
children, bees and, 28
chunk honey, 104
cleansing flights, 82
climate, 35–36
cluster, 44
Colony Collapse Disorder (CCD), 94
comb honey, 104
Commercial hive, 39
communication, bee methods for, 22
crown board. See inner cover
dances, honeybee, 22
deeps. See hive body
diseases, 85–86. See also individual diseases
division board, 48–49
drones, 19
dysentery. See nosema
eggs, honeybee, 21, 71
entrance reducer, 34, 35
essential oil treatment, 89
    See also Honey B-Healthy
European foulbrood, 85
evacuation, hive, 101–103
extracted honey, 104
extraction, honey, 37, 80, 98–103, 106–8
    equipment, 100
    timing of, 98–99
extractor, 100, 107–8
feeder pail, 49
feeders, 48–49
feeding bees, supplemental, 42–49, 75
    dry sugar, 45
    fondant, 45–46
    honey, 44–45
    pollen, 46–47
    sugar syrup, 45
    what to feed, 44–45
    when to feed, 44
filter, honey, 100
flowering plants. See bee garden
foam, in honey, 110

fondant: as food for bees, 45
    recipe, 46
foundation, 37
frame feeder. See division board
frame grip, 55
frame holder, 54–55
frames, 36–37
    removing, 69–70
    replacing, 72
Fruit, Nut, & Honey Granola (recipe), 125
fume board, 103
genetic diversity, 79
gloves, 53
Grain Mustard Honey Vinaigrette (recipe), 118
granola (recipe), 125
grease patties, for mite control, 89
grubs. See larva
haemolymph (bee blood), 16
hive, 18, 33
    accessibility of, 41
    purchasing established, 64–65
    siting requirements, 40–41
    sun exposure needs, 40
    types of, 39
    weight of, 36
hive body, 34, 35–36
    reversing, 76, 78
hive inspection, 66–73
    do's and don'ts, 67
    what to look for, 71
hive stand, 33–35
hive tool, 53, 69
hive-top feeder, 48, 68
Holiday Rounds (recipe), 116–17
honey, 96–112
    annual yield, 97
    as food for bees, 43, 44–45
    containers for, 109–10
    extraction, 37, 80, 98–103, 106–8
    flower source of, 110–11
    harvesting, ancient techniques for, 13
    infused (recipe), 115
    making of, 97
    nutritional value of, 112

selling, 110
storage of, 109
styles of, 104
Honey & Ginger Cold-Fighting Tea (recipe), 127
Honey B-Healthy, 88
honey flow. *See* nectar flow
honey house, 98. *See also* extraction, honey
Honey Ice Cream (recipe), 124
honey stores, 80
    assessing, 44, 81
    causes of diminished, 43
    recommendations, 44, 99
honeybee
    anatomy, 14–16
    antennae, 14
    cardiovascular system, 16
    developmental stages, 21
    digestion, 16
    diseases afflicting, 85–86.
    See also individual diseases
    eggs, 21
    eyes, 14
    honey stomach, 97
    larva, 21
    legs, 14
    proposcis, 14
    pupa, 21
    reproduction, 16, 18, 19, 22
    respiration, 16
    scent glands, 16
    sensory organs, 14
    species, 30–31
    stinger, 16
    wax glands, 16
    wings, 14
honeycomb: making of, 16, 33, 37
Honeyed Prawns & Polenta (recipe), 122–23
Hot Cider & Honey Toddy (recipe), 126
housing, 32–39
    ancient types, 13
    suppliers, 41
hygienic queens, 85, 88
ice cream, honey (recipe), 124
Infused Honey (recipe), 115

inner cover, 34, 37–38
integrated pest management (IPM), 85, 88
Kenya hive. *See* top bar hive
Langstroth hive, 13, 33
larva, 21, 71
legal considerations, 27
mice, 91
Miller feeder. *See* hive-top feeder
moisture, as threat to hive, 41
mouse guard, 91
Muth jars, 109–10
Nasonov pheromone, 22
National hive, 39
nectar, 43, 97
    bees gathering, 14, 97
nectar flow, 58, 75
neighborliness, 27
nosema, 86
nucs, 62
    making, 76–77
nutritional value, bee products, 112
obtaining bees, 56–65
    best time of year, 57
    suppliers, 65
opening a hive, 68
opossums, 91–92
organic certification, 111
outer cover, 34, 38
package bees, 57–61
    installing, 59–61
parasites, 87–89. *See also* individual parasites
pests, 90–93. *See also* individual pests
pets, bees and, 28
pheromones, 22
    queen's use of, 18, 22
    smoker's disruption of, 51
plastic bag feeder, 49
pollen, 43
    as food for bees, 46–47
    as food for humans, 112
pollen analysis, 110
pollen cakes, recipe, 47
pollen stores, assessing, 47
pollen substitutes, 47

pollination, 14, 43, 97
pollinators, 43
powdered sugar shake, for mite control, 87, 88
prawns, honeyed (recipe), 122–23
profiles: Chase, 64
    Corky, 38
    Debra, 83
    Eli, 113
    Jenny, 23
    John and Kate, 105
    Jon, 67
    Majora, 65
    Megan, 29
propolis, 14, 20
    as food for humans, 112
protective clothing, 28, 52–53
pupae, 21, 71
queen bee, 18–19
    finding, 18
    hygienic varieties, 85
    inspecting, 71
queen cage, 60–61
queen cell, 77
queen excluder, 34, 37, 54
queenless hives, 79
raccoons, 92
range, foraging, 97
recipes using honey, 114–27
repellants, bee, 103
reproduction, 16
requeening, 77, 79
Roasted Root Vegetables with Honey & Herbs
    (recipe), 119
robbing, 30
royal jelly, 18, 21
    as food for humans, 112
Sacbrood, 86
screened bottom board, 88
    See also bottom board
seasons. *See* spring, summer, autumn, winter
shipping bees, 57
site selection, 25, 40–41
skeps, 13, 39
skunks, 91–92

Index (continued)

small hive beetle, 90
Small Measure (blog), 10, 93
smoker, 22, 28, 51
 fuel for, 51
 instructions for use, 51–52, 68–70
smoking bees, 49
soup, chestnut (recipe), 120–21
space requirements, 25
species, honeybee, 30–31
splits, making, 76–77, 78
spring, beekeeping tasks during, 75–79
 checklist, 75, 128
stings: allergy to, 29
avoiding, 28
sugar candy. *See* fondant
sugar syrup, as food for bees, 45
sugar, dry, as food for bees, 45
summer, beekeeping tasks during, 80
 checklist, 80, 128
supers, 33, 34, 36
 storage of, 109
supersedure cells, 71, 79
supersedure, 18, 77, 79
suppliers, bees, 65, 129
swarm cells, 71,
removing, 78
swarm collection, 62–63
swarming, 27, 62–63, 79
 preventing, 78
tar paper wrapping, 82
tea (recipe), 127
top bar hive, 39, 70
tracheal mites, 89
uncapping comb, 100, 107
uncapping knife, 100, 106
varroa mites, 35, 75, 80, 87
 detecting, 88–89
 resistance to, 88
veil. *See* protective clothing
ventilation, 41, 78
vinaigrette (recipe), 118
water, hive needs for, 40, 78
waterer, hive, 48
wax cappings. *See* beeswax
wax cappings basin, 100, 106
wax glands, 16
wax moths, 90–91
WBC hive, 39

whipped honey, 104
windbreaks, for hives, 40
winter, beekeeping tasks during, 82
 checklist, 82, 128
woodenware, 33–38
painting, 41
workers, 19–20
 duties of, 20
wound care, honey for, 112